**Monographs on
Statistics and Applied Probability**

General Editors: **D. R. Cox and D. V. Hinkley**

Statistical Inference

Statistical Inference

S. D. Silvey
Formerly of University of Glasgow

Chapman and Hall
London   New York

First published 1970
by Penguin Books Ltd

Reprinted with corrections 1975
by Chapman and Hall Ltd
11 New Fetter Lane, London EC4P 4EE
Reprinted 1978, 1979, 1983

Published in the USA by
Chapman and Hall,
733 Third Avenue, New York, NY 10017

Printed in Great Britain at the
University Press, Cambridge

ISBN 0 412 13820 4

© 1975   S. D. Silvey

All rights reserved. No part of this publication may be
reproduced, stored in a retrieval system or transmitted in
any form or by any means, electronic, mechanical, photocopying,
recording, or otherwise, without the prior written permission
of the publisher.
This book is sold subject to the condition that it shall not,
by way of trade or otherwise, be lent, resold, hired out, or
otherwise circulated without the publisher's prior consent in
any form of binding or cover other than that in which it is
published and without a similar condition including this
condition being imposed on the subsequent purchaser.

Contents

Preface 11

1 Introduction 13
1.1 Preliminaries 13
1.2 The general inference problem 16
1.3 Estimation 18
1.4 Hypothesis testing 19
1.5 Decision theory 19
    Examples 1 19

2 Minimum-Variance Unbiased Estimation 21
2.1 The point estimation problem 21
2.2 'Good' estimates 22
2.3 Sufficiency 25
2.4 The Rao–Blackwell theorem 28
2.5 Completeness 29
2.6 Completeness and M.V.U.E.s 33
2.7 Discussion 34
2.8 Efficiency of unbiased estimators 35
2.9 Fisher's information 37
2.10 The Cramér-Rao lower bound 38
2.11 Efficiency 39
2.12 Generalization of the Cramér-Rao inequality 41
2.13 Concluding remarks 43
     Examples 2 44·

3 The Method of Least Squares 46
3.1 Examples 46
3.2 Normal equations 47
3.3 Geometric interpretation 48
3.4 Identifiability 50
3.5 The Gauss–Markov theorem 51
3.6 Weighted least squares 54
3.7 Estimation of σ^2 56

3.8 Variance of least-squares estimators 57
3.9 Normal theory 58
3.10 Least squares with side conditions 59
3.11 Discussion 64
Examples 3 65

4 The Method of Maximum Likelihood 68

4.1 The likelihood function 68
4.2 Calculation of maximum-likelihood estimates 70
4.3 Optimal properties of maximum-likelihood estimators 73
4.4 Large-sample properties 74
4.5 Consistency 76
4.6 Large-sample efficiency 77
4.7 Restricted maximum-likelihood estimates 79
Examples 4 84

5 Confidence Sets 87

5.1 Confidence interval 87
5.2 General definition of a confidence set 88
5.3 Construction of confidence sets 89
5.4 Optimal confidence sets 92
Examples 5 92

6 Hypothesis Testing 94

6.1 The Neyman–Pearson theory 96
6.2 Simple hypotheses 97
6.3 Composite hypotheses 102
6.4 Unbiased and invariant tests 104
Examples 6 106

7 The Likelihood-Ratio Test and Alternative 'Large-Sample' Equivalents of it 108

7.1 The likelihood-ratio test 108
7.2 The large-sample distribution of λ 112
7.3 The W-test 115
7.4 The χ^2 test 118
Examples 7 121

8 Sequential Tests 123

8.1 Definition of a sequential probability ratio test 124
8.2 Error probabilities and the constants A and B 125
8.3 Graphical procedure for an s.p.r. test 127
8.4 Composite hypotheses 129
8.5 Monotone likelihood ratio and the s.p.r. test 130
Examples 8 136

9 Non-Parametric Methods 139

9.1 The Kolmogorov–Smirnov test 140
9.2 The χ^2 goodness-of-fit test 142
9.3 The Wilcoxon test 143
9.4 Permutation tests 144
9.5 The use of a sufficient statistic for test construction 145
9.6 Randomization 148
Examples 9 151

10 The Bayesian Approach 153

10.1 Prior distributions 153
10.2 Posterior distributions 154
10.3 Bayesian confidence intervals 155
10.4 Bayesian inference regarding hypotheses 156
10.5 Choosing a prior distribution 157
10.6 Improper prior distributions 158
Examples 10 159

11 An Introduction to Decision Theory 161

11.1 The two-decision problem 161
11.2 Decision functions 162
11.3 The risk function 162
11.4 Minimax decision functions 165
11.5 Admissible decision functions 166
11.6 Bayes's solutions 166
11.7 A Bayes's sequential decision problem 171
Examples 11 175

Appendix A Some Matrix Results 177

Appendix B The Linear Hypothesis 180

References 189

Index 191

Preface

Statistics is a subject with a vast field of application, involving problems which vary widely in their character and complexity. However, in tackling these, we use a relatively small core of central ideas and methods. In this book I have attempted to concentrate attention on these ideas, to place them in a general setting and to illustrate them by relatively simple examples, avoiding wherever possible the extraneous difficulties of complicated mathematical manipulation.

In order to compress the central body of ideas into a small volume, it is necessary to assume a fair degree of mathematical sophistication on the part of the reader, and the book is intended for students of mathematics who are already accustomed to thinking in rather general terms about spaces, functions and so on. Primarily I had in mind final-year and postgraduate mathematics students.

Certain specific mathematical knowledge is assumed in addition to this general sophistication, in particular: a thorough grounding in probability theory and in the methods of probability calculus; a nodding acquaintance with measure theory; considerable knowledge of linear algebra, in terms of both matrices and linear transformations in finite-dimensional vector spaces; and a good working knowledge of calculus of several variables. Probability theory is absolutely essential throughout the book. However only parts of it require the other specific bits of knowledge referred to, and most of the ideas can be grasped without them.

There is a continuing controversy among statisticians about the foundations of statistical inference, between protagonists of the so-called frequentist and Bayesian schools of thought. While a single all-embracing theory has obvious attractions (and Bayesian theory is closer to this than frequentist theory), it remains true that ideas from both sides are useful in thinking about practical problems. So in this book I have adopted the attitude that I should include those ideas and methods which I have actually used in practice. At the same time, I have tried to present them in a way which encourages the reader to think critically about them and to form his own view of their relative strengths and weaknesses.

It is not my view that all that is required to make a statistician is an understanding of the ideas presented in this book. A necessary preliminary to their use in practice is the setting up of an appropriate probabilistic model for the situation under investigation, and this calls for considerable experience and judgement. I have made no attempt to discuss this aspect of the subject and the book contains no real data whatsoever. Moreover, the examples, some of which are not easy, are intended to provide the reader with an opportunity for testing his understanding of the ideas, and not to develop experience in establishing mathematical models. I consider it easier to grasp the basic ideas when one is not harassed by the necessity to exercise judgement regarding the model.

It is impossible for me to acknowledge individually my indebtedness to all those who have influenced my thinking about statistics, including present and past colleagues, but I must express my gratitude to three in particular: to Dr R. A. Robb who first introduced me to the subject and who supported me strongly yet unobtrusively during my early fumbling steps in applied statistics; to Professor D. V. Lindley whose lectures in Cambridge were most inspiring and whose strong advocacy of the Bayesian approach has forced many besides myself to think seriously about the foundations of the subject; and to Professor E. L. Lehmann whose book on testing statistical hypotheses clarified for me so many of the ideas of the frequentist school. I also wish to thank an anonymous referee for several suggestions which resulted in an improvement to an original version of the book.

Most of the examples for the reader are drawn from examination papers, and I am obliged in particular to the University of Cambridge for permission to reproduce a number of questions from papers for the Diploma in Mathematical Statistics. These are indicated by (*Camb. Dip.*). Since the original sources of questions are difficult to trace, I apologize to any colleague who recognizes an unacknowledged question of his own.

Finally I am extremely grateful to Miss Mary Nisbet who typed the manuscript with admirable accuracy and who sustained extraordinary good humour in the face of numerous alterations.

1 Introduction

1.1 Preliminaries

The theory of probability and statistics is concerned with situations in which an observation is made and which contain an element of inherent variability outwith the observer's control. He knows that if he repeated the observation under conditions which were identical in so far as he could control them, the second observation would not necessarily agree exactly with the first. Thus when a scientist repeats an experiment in a laboratory, no matter how careful he may be in ensuring that experimental conditions do not vary from one repetition to the next, there will be variation in the observations resulting from the different repetitions, variation which in this context is often referred to as experimental error. Similarly, two apparently identical animals will not react in exactly the same way to some given treatment. Indeed it is the case that almost all situations in which observations are taken contain this element of variability. So the field of application of the theory which will be introduced in this book is extremely wide.

The possibility of formulating a mathematical theory to assist in the interpretation of observations in this kind of situation arises from the following phenomenon. Suppose we consider an experiment which can be repeated and whose result is 'an observation'. This observation may belong to some stated set E of possible observations or it may not. If the observation belongs to E we shall say that 'the event E has occurred'. Now suppose that the experiment is repeated n times and on each occasion we note whether or not E occurs. It transpires in practice that the proportion of times that a stated event E occurs in n independent repetitions of the experiment seems to settle down, as n increases, to a fixed number, depending of course on E. This is the phenomenon which is popularly referred to as the 'law of averages' and it is this law which underlies the whole theory of probability. It leads to the description of the inherent variability in the kind of situation we are discussing by a *probability distribution* or *measure* over the set of possible observations.

It will be assumed that the reader is familiar with the notion of a probability distribution and the way in which the 'law of averages' motivates its underlying axioms. (For a full account of this he may consult such books as those by Lindgren, 1962, Lindley, 1965, Feller, 1968, Meyer, 1965.) However since there are slight differences in usage we shall now explain the interpretation to be given to certain terms which will occur repeatedly.

1.1.2 Sample spaces

Possible observations in a situation under investigation will be represented in a mathematical set or space called a *sample space*. It is not necessary that there be a one-to-one correspondence between elements of a sample space and possible observations. Indeed, in a sophisticated treatment an observation is represented by a set of points for reasons which we need not elaborate at this stage. It will be sufficient for our purposes, at least initially, to regard a sample space as a set in which each possible observation is represented by a distinct element or point. It would be unnecessarily restrictive and would lead to clumsiness to insist that *every* point in a sample space should represent a possible observation and so we shall allow the possibility that a sample space is bigger than is absolutely essential for the representation we have in mind. We shall denote this space by X, its typical point by x, and we shall refer to 'the observation x'. Of course X will vary according to the situation being investigated. It may be that each possible observation is a real number, in which case X may be taken as the set of real numbers. It happens frequently that each possible observation is an ordered set of n real numbers, in which case $x = (x_1, x_2, \ldots, x_n)$ and X may be taken as real n-space. It may even be that X is a space of functions, as, for example, when an observation is the curve traced by a barograph over a specified period of time.

1.1.3 Events

A subset E of a sample space X represents a real-life event, namely the event that the observation made belongs to the set of observations represented by E. Of course in everyday language a given event may not be described in this somewhat pedantic manner, but such a description of it is always possible. (To take an almost trivial example, if we consider rolling a die and observing the number on the face appearing uppermost, the event 'an even face turns up' is the event 'the observation made belongs to the set $\{2, 4, 6\}$ of possible observations'). Even if an event is described in everyday language it simplifies matters to think of it as simply a subset of a sample space, and because this habit of thought is so useful we shall often identify an event with the subset representing it and refer simply to 'the event E'.

We shall assume that the reader is familiar with the interpretation of the standard set operations in terms of events, for example, $E_1 \cup E_2$ is the event 'either E_1 or E_2'.

1.1.4 Probability distributions

As we have already said, the inherent variability in the kind of situation which concerns us is described by a probability distribution on the set of possible observations, or on a sample space. A useful, if somewhat naïve, way of thinking about a probability distribution is to imagine a unit mass spread over the sample space. The mass assigned thus to the subset E is the probability of the event E.

If the sample space X is discrete, that is, finite or countably infinite, a probability distribution on it is defined by stating the mass associated with each of its elements: then the probability of an event E is simply the sum of the masses associated with the elements of E. If X is not discrete and has some 'natural' measure such as length or area defined on it, then usually a probability distribution is defined by a probability density function relative to this 'natural' measure: then the probability of the event E is calculated by integrating this density function over E.

Thus if X is the real line, a probability density function is a non-negative function $p(x)$ of the real variable x, such that

$$\int_{-\infty}^{\infty} p(x)\,dx = 1.$$

The probability of the interval (a, b) is then

$$\int_a^b p(x)\,dx,$$

and these are ordinary Lebesgue (or Riemann) integrals. If X is the plane, a probability density function is a non-negative function $p(x_1, x_2)$ of the pair of real variables x_1 and x_2, such that

$$\int_{-\infty}^{\infty}\int_{-\infty}^{\infty} p(x_1, x_2)\,dx_1\,dx_2 = 1.$$

The probability, for instance, of the circle $C = \{(x_1, x_2) : x_1^2 + x_2^2 \leqslant 1\}$ is then

$$\iint_C p(x_1, x_2)\,dx_1\,dx_2,$$

and again these are ordinary Lebesgue double integrals.

In the sequel we shall usually be concerned with probability distributions which are defined in one or other of these ways. In both cases we shall write the probability $P(E)$ of the event E in the form

$$P(E) = \int_E p(x)\,dx.$$

In the discrete case this is to be interpreted as

$$\sum_{x_i \in E} p(x_i),$$

where $p(x_i)$ is the mass associated with x_i. In the non-discrete case dx is interpreted as the element of the 'natural' measure on X with respect to which

$p(x)$ is a density function, be it length, area etc., and the integral is interpreted accordingly. In both cases we shall refer to $p(x)$ as a (probability) density function.

1.1.5 *Random variables*

The phrase 'x is a (real) random variable with probability density function $p(x)$' is used with various shades of meaning by different authors. So far as we are concerned this phrase will be regarded simply as an alternative way of saying 'a sample space X for the situation of interest is the real line and the inherent variability of this situation is described by the probability distribution on X with density function $p(x)$.'

The reason for introducing this terminology is that it simplifies the description of more complex situations. Thus a very common statistical situation is the following. An experiment whose possible results are real numbers, is repeated n times, the replicates being independent of one another. Our typical observation x, then, is of the form $(x_1, x_2, \ldots, x_n)$ and the appropriate sample space is real n-space R^n. Moreover if the variability in the results of a single replicate is described by the probability distribution (on the line) with density function p, then that in the results of n independent replicates is described by the distribution (on n-space) with density function

$$f(x_1, x_2, \ldots, x_n) = \prod_{i=1}^{n} p(x_i).$$

This can all be conveyed simply by saying that $x_1, x_2, \ldots, x_n$ are independent random variables each with density function p.

An alternative way of describing the situation just enunciated is to say that $x_1, x_2, \ldots, x_n$ constitute a *random sample* from the distribution with density function p, and this phrase too, we shall sometimes use without going into the historical reasons for its use. Statistical literature still contains much quaint terminology like this.

1.2 **The general inference problem**

To define a probability distribution on a sample space it is not necessary to define the probability of *every* event. The distribution is completely defined by defining the probabilities of a sufficiently wide class of events, and then the rules or axioms of a probability distribution may be used to deduce the probabilities of events outside this class. It is with such deductions that probability calculus is concerned, with answering such questions as 'Given sufficient information to define a probability distribution completely, what is the probability of such-and-such an event?'

Statistical inference is concerned with a completely different kind of problem which arises for the following reasons. Suppose we consider an observational situation containing inherent variability. This is described by a probability distribution on a suitable sample space. However, in any specific case we shall not know what the appropriate probability distribution is. There will be a whole class of possible distributions one of which is appropriate for this case. Which one is unknown to us.

1.2.1 Trials

To take a specific example, suppose we consider a new model of automobile which is being produced in large numbers. We choose one at random from the production line and observe whether or not it suffers a mechanical breakdown within two years. This is the simplest kind of probabilistic situation, where there are only two possible observations. (Such a situation is often referred to as 'a trial'.) A suitable sample space consists of two elements 1 (representing breakdown) and 0 (representing no breakdown), and the inherent variability in the situation is described by a probability distribution which in this case is defined by a single number θ, the probability of breakdown. While we can say that the variability is described in this way, *for some value of* θ between 0 and 1, we do not know what this value is. In other words, there is a family $\{P_\theta : 0 \leq \theta \leq 1\}$ of possible distributions on the sample space and we do not know which one is appropriate.

1.2.2 Example

A more complicated illustration is the following. Suppose we have a large batch of seeds stored under constant conditions of temperature and humidity. In the course of time seeds die. Suppose that at time t a proportion $\pi(t)$ of the stored seeds are still alive. At each of times $t_1, t_2, \ldots, t_s$ we take a random sample of n seeds and observe how many are still alive. So a typical observation consists of an ordered set $(r_1, r_2, \ldots, r_s)$ of integers, r_i being the number of seeds observed to be alive at time t_i. If we know the function $\pi(t)$ then it is a standard result of probability calculus that the appropriate distribution for describing the variable element in this situation is that defined by the density

$$p(r_1, r_2, \ldots, r_s) = \prod_{i=1}^{s} \binom{n}{r_i} [\pi(t_i)]^{r_i} [1 - \pi(t_i)]^{n-r_i}.$$

Of course in practice $\pi(t)$ is unknown and so we have a class of possible distributions, one corresponding to each function $\pi(t)$. Now $\pi(t)$ is necessarily a non-increasing function of t, taking values between 0 and 1. If we denote the class of such functions by $\mathcal{D}$ and, as is customary, write θ instead of $\pi(t)$ to label distributions, again in this situation there is a family $\{P_\theta : \theta \in \mathcal{D}\}$ of possible distributions on the sample space and we do not know which is the appropriate, or true one.

The General Inference Problem

What has been illustrated in these two particular examples holds for a general observational situation containing inherent variability. There is a family $\{P_\theta : \theta \in \Theta\}$ of possible distributions describing this, and we do not know which is the true member of this family. The label θ is called a *parameter*. The set Θ to which the labelling parameter θ belongs varies from one situation to the next. In the first example above, Θ is a set of real numbers; in the second, it is a set of functions; in another example it might, for instance, be a set of vectors.

The general inference problem arises from the fact that we do not know which of a family of distributions is the true one for describing the variability of a situation in which we make an observation. From the observation made we wish to infer something about the true distribution, or equivalently about the true parameter. The general possibility of making inference rests in the fact that it is usually the case that a given observation is much more probable under some members of the family $\{P_\theta\}$ than it is under others, so that when this observation actually occurs in practice, it becomes plausible that the true distribution belongs to the former set rather than the latter. In this sense an observation gives information about the true distribution. We shall be concerned with clarifying this somewhat vague notion for particular types of problem and discussing the nature of the answers we can give.

1.3 Estimation

The first type of problem that we shall investigate is that of estimation.

The problem of *point estimation* arises when we are interested in some numerical characteristic of an unknown distribution (such as the mean or variance in the case of a distribution on the line) and we wish to calculate, from an observation, a number which, we infer, is an approximation to the numerical characteristic in question. Thus in example 1.2.2 we may be interested in the time t^* such that $\pi(t^*) = \frac{1}{2}$, and wish to use the observation $(r_1, r_2, \ldots, r_s)$ to calculate an approximation to, or estimate of, the unknown quantity t^*. There is little value in calculating an approximation to an unknown quantity without having some idea of how 'good' the approximation is and how it compares with other approximations. It is not immediately clear what we mean by 'good' in a probabilistic context where we have to *infer* that the calculated value approximates to the unknown quantity of interest. So we shall have to discuss what is meant by 'good' in this context.

As we have said, after an observation has been made, it becomes plausible that the true distribution belongs to a smaller family than was originally postulated as possible, or equivalently that the true value of the indexing parameter θ belongs to a proper subset of Θ. The problem of *set estimation* is concerned with determining such a plausible subset, and in clarifying the sense in which it is plausible.

1.4 Hypothesis testing

An observer sometimes has a theory which when translated into statistical language becomes a statement that the true distribution describing the inherent variability of his observational situation belongs to a smaller family than the postulated family $\{P_\theta : \theta \in \Theta\}$ of possible distributions, or equivalently that the true value of the parameter θ belongs to a subset of Θ. Thus in example 1.2.2, a botanist might have a theory explaining why seeds die which would imply a particular form for the survival function $\pi(t)$. In this case, he may wish to use an observation to infer whether or not his theory is true. It is with inferences of this nature that the theory of *hypothesis testing* is concerned, and this is the second of the main topics that we will discuss.

1.5 Decision theory

Many applications of statistics are based on the theories of estimation and hypothesis testing mentioned in sections 1.3 and 1.4. However, as we shall see, these theories have certain unsatisfactory features, and a deeper understanding of the nature of the problems involved is obtained by the study of decision theory, which is the last of our three main topics.

Revision examples on distribution calculus

1.1 Let $x_1, x_2, \ldots, x_n$ be independent $N(\mu, \sigma^2)$ random variables and let

$$\bar{x} = \frac{1}{n}(x_1 + x_2 + \ldots + x_n), \qquad s^2 = \frac{1}{n-1}\sum(x_i - \bar{x})^2.$$

Prove that
(a) $\bar{x}$ is $N(\mu, \sigma^2/n)$;
(b) $(n-1)s^2/\sigma^2$ is distributed as χ^2 with $n-1$ degrees of freedom;
(c) $\sqrt{n}(\bar{x}-\mu)/s$ is distributed as Student's t with $n-1$ degrees of freedom.

1.2 Let u and v be independently distributed as χ^2 with m and n degrees of freedom respectively. Show that nu/mv has an F-distribution.

1.3 Let $x_1, x_2, \ldots, x_n$ be independent Poisson random variables with common mean λ. Find the conditional distribution of x_1, given $x_1 + x_2 + \ldots + x_n$.

1.4 Let $x_1, x_2, \ldots, x_n$ be a random sample from the exponential distribution with density $e^{-u}(u > 0)$. Find the distribution of $\sum_{i=1}^{n} x_i$, and the conditional distribution of x_1, given $\sum_{i=1}^{n} x_i$.

1.5 If x is a random n-vector which is normally distributed with zero mean and variance matrix Σ, show that $x'\Sigma^{-1}x$ is distributed as χ^2 with n degrees of freedom. If $E(x) = \mu \neq 0$, what then is the distribution of $x'\Sigma^{-1}x$?

1.6 The random n-vector x is partitioned into $\begin{bmatrix} x_1 \\ x_2 \end{bmatrix}$ and its mean vector μ and variance matrix Σ are similarly partitioned into $\begin{bmatrix} \mu_1 \\ \mu_2 \end{bmatrix}$ and $\begin{bmatrix} \Sigma_{11} & \Sigma_{12} \\ \Sigma'_{12} & \Sigma_{22} \end{bmatrix}$ respectively. If x is normally distributed, find the conditional distribution of x_1, given x_2.

1.7 Let x be a random n-vector whose components are independent normal random variables each with zero mean and unit variance; and let P be a symmetric idempotent matrix of rank $r < n$. Prove that $x'Px$ and $x'(I-P)x$ are independent and that each is distributed as χ^2. (Here I is the unit matrix of order n.)

More generally, if $P_1, P_2, \ldots, P_k$ are symmetric idempotent matrices such that $P_1 + P_2 + \ldots + P_k = I$, show that $x'P_1 x, x'P_2 x, \ldots, x'P_k x$ are independent and that each is distributed as χ^2.

2 Minimum-Variance Unbiased Estimation

2.1 The point estimation problem

Before formulating the point estimation problem in a general mathematical way, we shall consider some particular examples of it.

2.1.1 If a situation is such that only two outcomes, often called success and failure, are possible, it is usually called a *trial*. The variable element in a trial is described by a probability distribution on a sample space of two elements, 0 representing failure and 1 success; this distribution assigning the probability $1-\theta$ to 0 and θ to 1, where $0 \leqslant \theta \leqslant 1$. Suppose we consider n independent repetitions of a given trial. The variable element in these is described by a probability distribution on a sample space of 2^n points, the typical point being $x = (x_1, x_2, \ldots, x_n)$, where each x_i is 0 or 1, and x_i represents the result of the ith trial. The appropriate probability distribution is defined by

$$p_\theta(x) = \theta^{m(x)} (1-\theta)^{n-m(x)},$$

where $m(x) = \sum_{i=1}^{n} x_i$ is the number of 1s in the results of the n trials, this being so since the trials are independent.

Given an x in this situation it seems reasonable to estimate θ by $m(x)/n$, the proportion of successes obtained. This seems in some sense to be a 'good' estimate of θ. We shall inquire in the course of this chapter in precisely what sense it is good.

2.1.2 A plausible probabilistic model of the way in which particles are emitted from a radioactive source yields the result that the number of particles emitted in a unit interval of time, may be regarded as a random variable with a Poisson distribution, that is,

$$\Pr(r \text{ particles are emitted in unit interval}) = \frac{e^{-\lambda} \lambda^r}{r!}.$$

(See, for instance, Lindley, 1965, vol. 1, section 2.3.)

For a given source, λ will in general be unknown and a scientist might be interested in the question, 'What is the emission rate λ of this source?' To answer this question he would observe the number N, say, of particles emitted by the source over some longish period of time T and call N/T the emission rate of the source. It is not difficult to see that this is an estimation problem, that the question is one of estimating λ, and that N/T is in some sense a reasonable estimate.

2.1.3 The variability of the point of impact of a bullet fired at an infinite target may, in virtue of the central limit theorem, be described by a 'bivariate normal distribution'; that is, by a pair (x_1, x_2) of random variables with density function

$$p(x_1, x_2) = \frac{1}{2\pi|\Sigma|^{\frac{1}{2}}} \exp\{-\tfrac{1}{2}(x-\theta)'\Sigma^{-1}(x-\theta)\},$$

where x' is the vector (x_1, x_2), θ' a vector (θ_1, θ_2), and Σ a positive definite two-by-two matrix. Here x_1 and x_2 are interpreted as the abscissa and ordinate of the point of impact relative to horizontal and vertical axes meeting at the centre of the target.

For a given marksman using a new rifle, the vector θ and the matrix Σ will be unknown. The marksman may fire several shots independently at the target in order to determine the 'centre of impact'. Since for the above bivariate normal distribution θ_1 is the mean of x_1, and θ_2 that of x_2, the marksman's object may be interpreted as the estimation of θ_1 and θ_2.

2.1.4 Suppose that a *stimulus* can be applied to *subjects* at various levels. For example, the stimulus might be a drug, the subject an animal, and the level a dose of the drug. In general the probability of a *response* to the stimulus depends on the level s at which it is applied, and often it may be assumed that the probability $\theta(s)$ of response to level s is a non-decreasing function of s. An experimenter may well be interested in a certain aspect of this 'response curve', for example in the 99 per cent response level, that is, in the value s^* such that $\theta(s^*) = 0.99$. Given the responses of subjects to different levels of the stimulus, he is then facing a problem of estimating a real valued function s^* of the unknown parameter θ (which in this case happens to be a function).

2.2 'Good' estimates

2.2.1 In each of the examples in section 2.1 the problem is one of using observations to find a number (or numbers) which, we infer, is an approximation (are approximations) to a numerical characteristic (or characteristics) of an unknown distribution. As always, one wishes such approximations to be 'as good as possible', and in order to clarify what we mean by this phrase in an inference context we shall now formulate the problem mathematically, considering first on the case where we are interested in a single numerical characteristic.

The basic ingredients in a mathematical description of this problem are a sample space X, a family $\{P_\theta : \theta \in \Theta\}$ of probability distributions on X, and a real-valued function g on Θ (a real parameter g). There is a 'true' (though unknown) value $g(\theta)$ of this function corresponding to the true but unknown θ, and this is what we wish to estimate. To do so we must state a rule which tells us, for each x in X, what we should use as an approximation to $g(\theta)$: in other words we must define a real-valued function $\hat{g}$ on X. Such a function $\hat{g}$ is called an *estimator* of $g(\theta)$ and a particular value of it, $\hat{g}(x)$, is called an estimate. While it is useful to be aware of this distinction between estimate and estimator, it becomes somewhat pedantic to maintain it when it is clear from the context that we wish to regard $\hat{g}(x)$ as a function – as when we write $E\{\hat{g}(x)\}$. So we shall not maintain the distinction consistently in the sequel. What we now have to do is to say what we mean by a 'good' estimator.

Relative to any particular member of the family $\{P_\theta\}$ of distributions on X we can make probability statements about $\hat{g}$. (There is a technical point here. We must assume that the function $\hat{g}$ is measurable, and we shall invariably make this assumption in the sequel. This need not concern the reader who is unfamiliar with measure theory. However for an explanation of measurability and other relevant notions of measure theory, he is referred to Doob, 1953, supplement.)

Thus $\quad P_\theta\{|\hat{g}-g(\theta)| > c\} = P_\theta\{x : |\hat{g}(x)-g(\theta)| > c\}$

is a well defined mathematical expression whose practical interpretation is as follows. It is the probability that $\hat{g}$ differs from the 'true' value $g(\theta)$ of the numerical characteristic of interest by more than c. (When we speak of the 'true' distribution P_θ we simply mean that any probability statements made refer to the distribution P_θ on the sample space.)

2.2.2 Now ideally we should like to have

$P_\theta\{\hat{g} = g(\theta)\} = 1 \quad$ for every θ,

since the practical interpretation of this statement is, 'With probability 1 our approximation or estimator $\hat{g}$ equals the true parameter $g(\theta)$ whatever this may be.' There is a case where this can be achieved. Suppose that the sample space X can be partitioned into a family $\{X_\theta : \theta \in \Theta\}$ of (disjoint) subsets and that the distribution P_θ assigns probability 1 to X_θ for every θ. Then if

$\hat{g}(x) = g(\theta) \quad$ for $x \in X_\theta$ and every $\theta \in \Theta$,

the estimator $\hat{g}$ of $g(\theta)$ contains essentially no error. Of course from the practical point of view this is a situation which virtually never occurs, though, particularly when the observation x constitutes a large random sample from a distribution, it may be approached quite closely.

2.2.3 A less formal way of describing this ideal situation is to say that in it, any observation x is *possible* under only one member of the family $\{P_\theta\}$ of distributions, so that when we observe an x we *know* which is the true distribution, and hence what is the true value of the real parameter of interest. Usually every observation x is possible under every distribution. Then we cannot estimate without some possibility of error. So we are led to the more realistic consideration that a good estimator is one which in some average sense is as good an approximation to the parameter of interest as possible. One obvious demand of this kind to make is that an estimator should have minimum mean-square error, that is, that a good estimator $\hat{g}$ should be such that

$$E_\theta\{\hat{g}-g(\theta)\}^2 \leqslant E_\theta\{\tilde{g}-g(\theta)\}^2,$$

for every θ and every other estimator $\tilde{g}$.

Unfortunately, as a little thought shows, this is not a realistic demand either. For any given θ, θ_0 say, we can always find an estimator whose mean-square error is zero for this particular θ (the estimator defined by $\tilde{g}(x) = g(\theta_0)$ for all x). Thus to have, *uniformly in* θ, minimum mean-square error, an estimator $\tilde{g}$ must have zero mean-square error for *every* θ. That is, it must be possible to estimate $g(\theta)$ without essential error, and as we have seen this is usually impossible. So we must modify our demands further.

2.2.4 Unbiasedness

An estimator which takes the same value for all x is clearly ridiculous from the practical point of view. If we use such an estimator we might as well take no observation at all and merely state *a priori* that our estimate of $g(\theta)$ is so and so. Now it may be that if we start by throwing out such absurd estimators, there may exist one which has minimum mean-square error among the remainder. One way of eliminating these trivial estimators is to demand that an estimator be *unbiased*, that is, that its expected value should equal the true parameter value, whatever this may be, so that

$$E_\theta\{\hat{g}\} = g(\theta) \quad \text{for all } \theta.$$

It is to be noted that this demand is more artificial than any we have previously considered and that while it does eliminate trivial estimators, it may also eliminate quite respectable estimators. For instance, suppose that $x_1, x_2, \ldots, x_n$ is a random sample from a normal distribution with unknown mean μ and unknown variance σ^2. (The indexing parameter θ of this family is $\theta = (\mu, \sigma^2)$.) 'Natural' estimates of the real parameters μ and σ^2 are the sample mean $\bar{x}$ and the sample variance

$$s^2 = \frac{1}{n}\sum_{i=1}^{n}(x_i-\bar{x})^2.$$

Of these the first is unbiased, $E_\theta(\bar{x}) = \mu$, for all θ, but the second is biased since $E(s^2) = (n-1)n^{-1}\sigma^2$, for all θ. Despite this, however, a great deal of attention has been paid to unbiased estimates and we shall now consider the question, 'Within the class of unbiased estimates does there exist one of minimum mean-square error?' Of course, for an *unbiased* estimator, mean-square error,

$E_\theta\{\hat{g}-g(\theta)\}^2$, is the same thing as variance, $\text{var}_\theta \hat{g}$, so that we may talk about a minimum-variance unbiased estimator, or an M.V.U.E.

2.3 Sufficiency

The idea of sufficiency is an important notion in statistical theory generally as well as in minimum-variance unbiased estimation in particular, so that while we discuss it in the latter context it must be borne in mind that it has more general impact.

Intuitively it is fairly clear that sometimes certain aspects of a set x of observations provide no information at all about an unknown parameter θ of interest, and that in making inferences about θ, we lose nothing by neglecting such aspects. For example, if x denotes the results of n *independent* trials in each of which the probability of success is θ, so that $x = (x_1, x_2, \ldots, x_n)$, where each x_i is either 1 (for success) or 0 (for failure), then it is plausible that only the sum $\sum x_i$ (the number of successes) provides information about θ and that, for instance, the order in which the 0s and 1s occur is totally irrelevant and may be ignored in estimating θ. (It is important to bear in mind that these other aspects of the observation x may provide information about the validity of the model adopted, that is, in this case about whether the trials are indeed independent and identical. If we are not sure of this assumption, we may well use these other aspects to test its validity. However this is not our concern in the meantime.)

How can we give this notion general mathematical expression? This can be done in terms of partitions of the sample space.

Suppose that we have a partition $\mathscr{A}$ of the sample space, that is, a covering of the sample space by a family $\mathscr{A}$ of disjoint subsets, and we consider the conditional probability distribution $P_{\theta|A}$ corresponding to a parameter θ, over a set A of this partition. Suppose further that for no $A \in \mathscr{A}$ does $P_{\theta|A}$ depend on θ. Now imagine that we are given the set A of the partition $\mathscr{A}$ to which a point x in the same space belongs. This piece of information may give us some information about θ, but further information about which point of A x happens to be, gives information only about the conditional distribution $P_{\theta|A}$, and since this does not depend on θ, this further information tells us nothing more about θ.

Consider, for instance, the case of n independent trials. The typical point of the sample space is $x = (x_1, x_2, \ldots, x_n)$, where each x_i is either 0 or 1, and there are 2^n points in the sample space.

Let $\mathscr{A}$ be the partition $\{A_0, A_1, \ldots, A_n\}$, where $x \in A_r$ if and only if

$$\sum_{i=1}^n x_i = r.$$

There are $\binom{n}{r}$ points in A_r. If θ is the probability of success for each trial, then

$$P_{\theta|A_r}(x) = \binom{n}{r}^{-1}$$

for every $x \in A_r$ and every θ between 0 and 1. So the conditional distribution $P_{\theta|A_r}$ is independent of θ for every A_r, and this is a formal mathematical expression of the fact that, given the total number of successes, additional information concerning the order of occurrence of successes and failures tells us no more about θ.

We are then led to the following general definition.

2.3.1 *Definition*

A partition $\mathscr{A}$ of the sample space is said to be sufficient *for θ (more precisely, for the family $\{P_\theta : \theta \in \Theta\}$) if for no $A \in \mathscr{A}$ does $P_{\theta|A}$ depend on θ.*

It is immediately obvious that the partition whose elements are individual points of the sample space is sufficient for an unknown parameter and that this partition is 'finer' than any other sufficient partition, but it often happens that there is a 'coarser' partition than this which is also sufficient. It may be that there exists a sufficient partition $\mathscr{A}$ which is at least as coarse as any other sufficient partition $\mathscr{B}$ in the sense that every set of $\mathscr{B}$ is contained in a set of $\mathscr{A}$. Such a partition $\mathscr{A}$ is said to be *minimal sufficient*.

The question of the existence, in general, of a minimal-sufficient partition involves measure-theoretic difficulties. (For a discussion of this, see Lehmann and Scheffe, 1950.) These difficulties disappear when the sample space is either discrete or a finite-dimensional Euclidean space, and the family $\{P_\theta\}$ of distributions is defined by a family $\{p_\theta\}$ of density functions with respect to the 'natural' measure on the appropriate space. Then it is possible to construct a minimal-sufficient partition as follows.

Two points x and x' in the sample space are said to be equivalent and we write $x \sim x'$ if the ratio $p_\theta(x)/p_\theta(x')$ of densities does not depend on θ. This is an equivalence relation and so it defines a partition of the sample space. This partitition is minimal sufficient.

Consider, for instance, the case of n independent trials when

$x = (x_1, x_2, \ldots, x_n)$ and $p_\theta(x) = \theta^{\Sigma x_i}(1-\theta)^{n-\Sigma x_i}$.

Then $\dfrac{p_\theta(x)}{p_\theta(x')} = \theta^{\Sigma x_i - \Sigma x'_i}(1-\theta)^{-\Sigma x_i + \Sigma x'_i}$.

This ratio does not depend on θ iff $\sum x_i = \sum x'_i$. So $x \sim x'$ iff $\sum x_i = \sum x'_i$. Hence the partition $\mathscr{A} = \{A_0, A_1, \ldots, A_n\}$, where $x \in A_r$ iff $\sum x_i = r$, introduced above, is not only sufficient; it is minimal sufficient.

While partitions are useful as an aid to understanding the notion of sufficiency, this idea is usually expressed in terms of statistics. A *statistic* is simply a function (not necessarily real valued) on the sample space. To any such function t corresponds a partition of the sample space, the typical set A of this partition being defined as $t^{-1}(a)$, for a fixed a in the range of t. Conversely, given a partition of the sample space we can clearly define a function for which the corresponding partition is the given one, though of course this function is not, in general, unique.

2.3.2 Sufficient statistic

A *statistic* t is said to be sufficient for a parameter θ (more precisely, for the family $\{P_\theta : \theta \in \Theta\}$) if the partition of the sample space generated by t is sufficient or, in other words, if the conditional distribution $P_{\theta|t}$ on the sample space, for given t, does not depend on θ. Similarly t is said to be *minimal sufficient* if its corresponding partition is minimal sufficient, so that *a minimal-sufficient statistic is a function of every other sufficient statistic*. A minimal-sufficient statistic corresponds to the greatest reduction of a set of data (comprising an observation) that can be achieved without discarding any information relevant to inferences about the unknown distribution. It is worth noting that a minimal-sufficient statistic, if it exists, is not unique since any one-to-one function of a minimal-sufficient statistic is also minimal sufficient. (They generate the same partition of the sample space.)

How in practice do we recognize a sufficient statistic? In the case where the family $\{P_\theta\}$ of distributions on the sample space is defined by a family $\{p_\theta\}$ of density functions with respect to some fixed measure, a most useful result in this context is the following.

2.3.3 The factorization theorem

A necessary and sufficient condition for a statistic t to be sufficient for a family $\{P_\theta : \theta \in \Theta\}$ of distributions is that $p_\theta(x)$ can be expressed in the form

$$p_\theta(x) = g_\theta\{t(x)\} h(x),$$

where g_θ and h are appropriately measurable functions of the indicated variables and h does not depend on θ.

A complete proof of this result involves measure-theoretic considerations which are beyond the scope of this book. The interested reader may refer to Lehmann (1959), p. 47. In the case where the sample space is discrete these difficulties disappear and we now give a proof for this case. In this case $p_\theta(x)$ is the probability assigned to the point x by the distribution P_θ.

(i) Let $t(x) = a$ and suppose that the factorization criterion is satisfied so that $p_\theta(x) = g_\theta(a) h(x)$.

Then $P_\theta(t = a) = \sum_{x' \in t^{-1}(a)} p(x') = g_\theta(a) \sum_{x' \in t^{-1}(a)} h(x').$

Hence $P_\theta(x|t = a) = \dfrac{p_\theta(x)}{P_\theta(t = a)} = \dfrac{h(x)}{\sum_{x' \in t^{-1}(a)} h(x')}$,

and this does not depend on θ.

(ii) Conversely, if t is sufficient,

$$p_\theta(x) = P_\theta(t = a)P(x|t = a),$$

where the second factor does not depend on θ because of sufficiency. Writing $P_\theta(t = a) = g_\theta(a)$ and $P(x|t = a) = h(x)$ gives the result.

2.3.4 In the case of n independent trials with probability θ of success we have

$$p_\theta(x) = \theta^{\Sigma x_i}(1-\theta)^{n-\Sigma x_i}.$$

Trivially this factorizes into a function of θ and t, where $t(x) = \sum x_i$ and a function of x only – the function $h(x) \equiv 1$.

2.3.5 Let $x = (x_1, x_2, \ldots, x_n)$ be a random sample of n from a $N(\mu, \sigma^2)$ distribution with $\theta = (\mu, \sigma)$ unknown.

Then $p_\theta(x) = \dfrac{1}{(2\pi\sigma^2)^{\frac{1}{2}n}} \exp\left[-\dfrac{1}{2\sigma^2}\sum(x_i - \mu)^2\right]$

$= \dfrac{1}{(2\pi\sigma^2)^{\frac{1}{2}n}} \exp\left[-\dfrac{1}{2\sigma^2}\left\{\sum(x_i - \bar{x})^2 + n(\bar{x} - \mu)^2\right\}\right].$

Again trivially, $p_\theta(x)$ factorizes into a function of t, where $t(x) = (\bar{x}, \sum(x_i - \bar{x})^2)$ a vector-valued function, and θ and a function of x only (again the constant function 1). So t is sufficient for θ.

This example illustrates that if our observation is a random sample of n from a normal distribution, we lose no information relevant to inferences by quoting merely the sample mean and the sample variance. Note, however, that this is not usually true if the family of possible distributions is other than normal.

2.4 The Rao–Blackwell theorem

The relevance of sufficiency to the question of minimum variance unbiased estimators is brought out by the following theorem.

2.4.1 *Let $\{P_\theta : \theta \in \Theta\}$ be a family of distributions on a sample space X and suppose that $\tilde{g}$ is an unbiased estimator of a real-valued function g of θ. Then if t is a sufficient statistic for θ, $E_\theta\{\tilde{g}|t\}$ is also an unbiased estimator of g and it has variance uniformly no larger than that of $\tilde{g}$.*

Proof. We have to prove three things:
(i) that $E_\theta\{\tilde{g}|t\}$ is a function $\hat{g}(t)$, say, of t only and does not depend on θ;
(ii) that $E_\theta\{\hat{g}(t)\} = g(\theta)$ for all $\theta \in \Theta$;

(iii) that $\text{var}_\theta(\hat{g}) \leq \text{var}_\theta(\tilde{g})$ for all $\theta \in \Theta$.

Now (i) is true because, since t is sufficient for θ, the conditional θ-distribution on X, given t, and therefore the conditional θ-expectation of any function on X, given t, does not depend on θ.

Also (ii) is true because

$$g(\theta) = E_\theta(\tilde{g}) = E_\theta\{E_\theta(\tilde{g}|t)\} = E_\theta\{\hat{g}(t)\}.$$

Finally (iii) is a particular case of the following more general result. Let c be a continuous convex function of a real variable. Then Jensen's inequality – which is little more than the statement that

$$c\{\lambda u_1 + (1-\lambda)u_2\} \leq \lambda c(u_1) + (1-\lambda)c(u_2)$$

– states that if u is a real random variable $c\{E(u)\} \leq E\{c(u)\}$. This inequality holds also for conditional expectations (see Doob, 1953, p. 33). Now, by a standard result on iterated expectations (see Meyer, 1965, p. 135), we have

$$E_\theta\{c(\tilde{g})\} = E_\theta[E_\theta\{c(\tilde{g})|t\}].$$

By Jensen's inequality,

$$E_\theta[E_\theta\{c(\tilde{g})|t\}] \geq E_\theta[c\{E_\theta(\tilde{g}|t)\}]$$
$$= E_\theta[c\{\hat{g}(t)\}].$$

If we now set $c(u) = \{u - g(\theta)\}^2$ we have the result that

$$\text{var}_\theta(\tilde{g}) \geq \text{var}_\theta(\hat{g}).$$

2.4.2 Before this theorem can be used to establish the existence of M.V.U.E.s, we have to demand the existence of a sufficient statistic t with an additional property. Suppose that we are estimating $g(\theta)$ and that there exists a sufficient statistic t with the property that there is a *unique* function of t, $\hat{g}(t)$ say, which is an unbiased estimator of $g(\theta)$. Then $\hat{g}(t)$ is an M.V.U.E. of $g(\theta)$. For if $\tilde{g}(x)$ is *any other* unbiased estimate of $g(\theta)$, $E\{\tilde{g}(x)|t\}$ must be $\hat{g}(t)$; because it is a function of t which is an unbiased estimator. Consequently, by the Rao–Blackwell theorem

$$\text{var}_\theta\{\hat{g}(t)\} \leq \text{var}_\theta\{\tilde{g}(x)\}.$$

Since $\tilde{g}$ is *any* unbiased estimator it follows that $\hat{g}(t)$ is M.V.U.

So another question arises. How can we ascertain whether a sufficient statistic has this additional property? Sometimes we can do so by using the notion of *completeness* of a family of distributions.

2.5 Completeness

Definition. Let $\{P_\theta^y : \theta \in \Theta\}$ *be a family of distributions on a space* Y *of points* y. *This family is said to be* complete *if* $E_\theta\{f(y)\} = 0$ *for all* $\theta \in \Theta$ *implies*

$f(y) = 0$ almost everywhere. If in this statement f is restricted to be a bounded function, the family is said to be *boundedly complete*. *Clearly completeness implies bounded completeness.*

2.5.1 It is comparatively easy to see that if a family of distributions on a space Y is boundedly complete then no non-trivial partition of the space is sufficient for the family. Suppose, for instance, that each set of a sufficient partition of Y contains just two points: let the points of a typical set A of the partition be y_1, y_2. The conditional distribution $P_{\theta|A}$ over these points does not depend on θ. Suppose that the probabilities assigned to y_1 and y_2 by this conditional distribution are

p and $1-p$ respectively. Let $f(y_1) = \dfrac{1}{p}$ and $f(y_2) = -\dfrac{1}{1-p}$.

Then $E(f|A) = E_\theta(f|A) = 0$.

Defining f similarly for every element A of the partition yields a function, not identically zero, such that $E_\theta(f) \equiv 0$, so that the family of distributions is not boundedly complete. The general case, where each element of the partition $\mathscr{A}$ may have more than two points is similar.

It follows that completeness is related to minimal sufficiency. Suppose that t is a sufficient statistic for a family $\{P_\theta : \theta \in \Theta\}$ of distributions on a sample space X and that the family $\{P_\theta^t : \theta \in \Theta\}$ *of distributions of t is boundedly complete.* Then no function of t which generates a coarser partition of the sample space then that generated by t is sufficient, by the previous argument. That is, only one-to-one functions of t are sufficient. Subject then to mild conditions which ensure the existence of a minimal-sufficient statistic, t is minimal sufficient. However, as we shall see, it is *not* true that the family of distributions of a minimal-sufficient statistic is necessarily complete.

2.5.2 *Example*

Consider the family of binomial distributions with densities

$$\binom{n}{t} \theta^t (1-\theta)^{n-t} \qquad 0 \leqslant \theta \leqslant 1 \qquad t = 0, 1, \ldots, n$$

and suppose that

$$\sum_{t=0}^{n} f(t) \binom{n}{t} \theta^t (1-\theta)^{n-t} = 0 \quad \text{for all } \theta \text{ in } [0,1].$$

This implies that

$$\sum_{t=0}^{n} f(t) \binom{n}{t} \left(\frac{\theta}{1-\theta}\right)^t = 0 \quad \text{for all } \theta \text{ in } (0,1),$$

that is, $\sum_{t=0}^{n} f(t) \binom{n}{t} \phi^t = 0 \quad \text{for all } \phi > 0.$

This polynomial identity implies that $f(t) = 0$, $t = 0, 1, \ldots, n$ and so the family of binomial distributions is complete.

Now suppose that we are dealing with n independent trials with probability θ of success. We have seen that the statistic t defined as the number of successes obtained is sufficient for θ. The family of distributions of t (varying θ) is the binomial family just discussed. This family is complete, hence in particular is boundedly complete.

Example 2.5.2 is a particular case of a useful general result on completeness of an important kind of family of distributions, the so-called exponential family.

2.5.3 Definition

A family of distributions on a Euclidean space is said to belong to the exponential family if it is defined by density functions, with respect to some fixed measure μ, of the form

$$p_\theta(x) = C(\theta) \exp\left[\sum_{i=1}^{k} Q_i(\theta) t_i(x)\right] h(x),$$

where θ is some parameter (not necessarily real valued), and the Qs and ts are real-valued functions.

The exponential family includes many classes of distributions which arise in practice; for example, binomial, normal, geometric, exponential, Poisson. The completeness theorem for this family is concerned with the case where θ varies over a subset Θ of k-dimensional space; and if $\theta = (\theta_1, \theta_2, \ldots, \theta_k)$, $Q_i(\theta) = \theta_i$.

2.5.4 Completeness theorem for an exponential family

Let an exponential family of distributions be defined by the density functions (with respect to a σ-finite measure μ)

$$p_\theta(x) = C(\theta) \exp\left[\sum_{i=1}^{k} \theta_i t_i(x)\right] \qquad \theta \in \Theta$$

and let $t(x) = \{t_1(x), t_2(x), \ldots, t_k(x)\}$. Then the family $\{P_\theta^t : \theta \in \Theta\}$ of distributions of t is complete if Θ contains a k-dimensional rectangle.

Again for a proof of this result the reader is referred to Lehmann (1959), p. 132.

2.5.5 Example

Let $x = (x_1, x_2, \ldots, x_n)$ be a random sample from an $N(\mu, \sigma^2)$ distribution, so that if $\theta = (\mu, \sigma^2)$

$$p_\theta(x) = \frac{1}{(2\pi)^{\frac{1}{2}n}\sigma^n} \exp\left[-\frac{1}{2\sigma^2}\sum(x_i-\mu)^2\right]$$

$$= \frac{1}{(2\pi)^{\frac{1}{2}n}\sigma^n} \exp\left[-\frac{1}{2\sigma^2}\left(\sum x_i^2 - 2\mu\sum x_i + n\mu^2\right)\right]$$

$$= \frac{\exp(-n\mu^2/2\sigma^2)}{(2\pi)^{\frac{1}{2}n}\sigma^n} \exp\left[-\frac{\sum x_i^2}{2\sigma^2} + \frac{\mu}{\sigma^2}\sum x_i\right]$$

$$= C(\theta)\exp\{Q_1(\theta)t_1(x) + Q_2(\theta)t_2(x)\},$$

where $C(\theta) = (2\pi)^{-\frac{1}{2}n}\sigma^{-n}\exp\left[-\frac{n\mu^2}{2\sigma^2}\right]$,

$Q_1(\theta) = -\frac{1}{2\sigma^2} \qquad t_1(x) = \sum x_i^2$,

$Q_2(\theta) = \frac{\mu}{\sigma^2} \qquad t_2(x) = \sum x_i$.

So we are dealing with an exponential family on $\mathbb{R}^n$.

Now let us 'reparametrize' by setting $\theta_1^* = -1/2\sigma^2$, $\theta_2^* = \mu/\sigma^2$, $\theta^* = (\theta_1^*, \theta_2^*)$.

This establishes a one-to-one correspondence between possible values of θ and values of θ^*. Obviously we may label the family of distribution on the sample space by θ^* rather than by θ, without altering the family in any way. We have

$$p_\theta^*(x) = C^*(\theta^*)\exp\{\theta_1^* t_1(x) + \theta_2^* t_2(x)\}.$$

If the original parameter space contains a two-dimensional rectangle, (for instance, as is usually the case, it is a half-plane), then so does the new parameter space, and it follows from theorem 2.5.4 that the family of distributions of $t = (t_1, t_2)$ is complete.

Note, however, that this is not necessarily so if the original parameter space does not contain a two-dimensional rectangle. Suppose, for instance, that $\Theta = \{\theta : \mu = \sigma^2\}$. Then $\Theta^* = \{\theta^* : \theta_1^* < 0, \theta_2^* = 1\}$ and Θ^* does not contain a two-dimensional rectangle. We cannot then use theorem 2.5.4 to deduce completeness of the family $\{P_\theta^t : \theta \in \Theta\}$.

The smaller the size of a family of distributions, the easier it is to find a function which is not identically zero, but whose expectation is zero for every member of the family. In the example above, we are reducing the size of the family of possible normal distributions when we limit consideration to those for which $\mu = \sigma^2$. For each member of this limited family, the expectation of

$$f(x) = \bar{x} - \frac{1}{n-1}\sum(x_i - \bar{x})^2$$

is zero, while $f(x)$ is obviously not identically zero. So the limited family is not complete. On the other hand, the expectation of $f(x)$ is not zero for every member of the larger family, unrestricted by the condition that $\mu = \sigma^2$. And indeed the larger family is complete, as we have seen.

2.6 Completeness and M.V.U.E.s

The use of completeness in connexion with M.V.U.E.s is straightforward. For suppose that t is a sufficient statistic whose family of distributions is complete. Then if there exists a function $\hat{g}(t)$ which is an unbiased estimator of a real parameter $g(\theta)$, $\hat{g}(t)$ is unique in this respect. For if $\hat{g}(t)$ and $\overset{\circ}{g}(t)$ are two such functions we have

$$E_\theta\{\hat{g}(t) - \overset{\circ}{g}(t)\} = g(\theta) - g(\theta) = 0 \quad \text{for every } \theta.$$

Hence by completeness $\hat{g}(t) \equiv \overset{\circ}{g}(t)$.

So, by the argument of section 2.4.2, $\hat{g}(t)$ is an M.V.U.E. of $g(\theta)$.

2.6.1 Example

Suppose that $x = (x_1, x_2, \ldots, x_n)$ is a random sample from the distribution on the positive numbers with density $\theta^2 x\, e^{-\theta x}$, $\theta > 0$. Determine a minimum variance unbiased estimator of θ.

We have density functions with respect to Lebesgue measure and

$$p_\theta(x) = \theta^{2n} \exp(-\theta \sum x_i) \prod_{i=1}^n x_i.$$

From this expression for the density function, two points emerge immediately:

(i) The statistic t defined by $t(x) = \sum x_i$ is sufficient for θ by the factorization theorem.

(ii) Because we are dealing with an exponential family containing a single unknown parameter and the range of this parameter certainly contains an interval on the line – the range is $(0, \infty)$ – the family $\{P_\theta^t : 0 < \theta < \infty\}$ of distributions of t is complete.

Hence the problem is solved if we can find an unbiased estimator $\tilde{\theta}$ of θ and then determine $E(\tilde{\theta}|t)$. It is not difficult to spot that $1/x_1$ is an unbiased estimator of θ. Consequently we now have to determine only $E(1/x_1|t)$.

In order to do this we first find the joint distribution of x_1 and t. Omitting details, we find that this joint distribution has density

$$\frac{\theta^{2n}}{\Gamma(2n-2)} x_1 (t-x_1)^{2n-3} e^{-\theta t} \quad t > 0 \quad 0 \leqslant x_1 \leqslant t.$$

From this it follows that the density of the conditional distribution of x_1, given t, is

$$(2n-1)(2n-2)\frac{x_1}{t^2}\left(1-\frac{x_1}{t}\right)^{2n-3} \qquad 0 \leqslant x_1 \leqslant t,$$

and that, in particular,

$$E\left(\frac{1}{x_1}\bigg|t\right) = \frac{2n-1}{t}.$$

So the required M.V.U.E. of θ is

$$\frac{2n-1}{\sum_{i=1}^{n} x_i}.$$

It is worth remarking that judicious guesswork may well be used to reduce the calculation in this type of example. For instance, it is possible in the example just considered to spot that $E(1/\sum x_i)$ has the form $a\theta$. Then we need only evaluate a, and this does not involve considering conditional distributions at all; it requires only the unconditional distribution of $\sum x_i$. Therefore evaluation of a is relatively easy. In this way we find more directly the M.V.U.E., namely $(2n-1)/\sum x_i$.

2.7 Discussion

To what extent we have solved the problem of the existence of M.V.U.E.s and of their determination when they exist, depends on what we can say about the existence of a sufficient statistic whose family of distributions is complete. Now there are many instances where the family of distributions of a minimal-sufficient statistic is not complete. The following is a simple example.

2.7.1 *Example*

Suppose that n independent trials each with probability θ of success are carried out; trials are then continued until an additional k successes are obtained, this requiring s additional trials. Let the results be denoted by $x = (x_1, x_2, \ldots, x_n, x_{n+1}, \ldots, x_{n+s-1}, 1)$ where as usual each x_i is either 1 (success) or 0 (failure).

We have

$$p_\theta(x) = \theta^{r+k}(1-\theta)^{n+s-r-k},$$

where $r = \sum_{i=1}^{n} x_i$; and s depends on x also.

It is clear from the factorization theorem that the statistic $t = (r, s)$ is sufficient for θ. Moreover, in obvious notation,

$$\frac{p_\theta(x)}{p_\theta(x')} = \left(\frac{\theta}{1-\theta}\right)^{r-r'} (1-\theta)^{s-s'}$$

and this does not depend on θ if and only if $r = r'$ and $s = s'$.

So the statistic t is minimal sufficient; but the family of distributions of t is *not* complete because,

if $\quad f(t) = \dfrac{r}{n} - \dfrac{k-1}{s-1}$,

$E_\theta\{f(t)\} = 0 \quad$ for every θ in $(0, 1)$,

and $f(t) \not\equiv 0$.

Lack of completeness of the family of distributions of a minimal sufficient statistic prevents our using the Rao–Blackwell theorem to establish the existence of an M.V.U.E. of a real parameter. For then there may be several different functions of the minimal-sufficient statistic which are unbiased estimators of the parameter and we have no general means of comparing their variances. We have therefore only very partially solved the problem of minimum variance unbiased estimation.

2.8 Efficiency of unbiased estimators

While it is desirable from a theoretical point of view to demonstrate that an estimator is, in some sense, the best possible, it may be sufficient from a practical point of view to show that some estimator which is being used is nearly the best possible. If, somewhat arbitrarily, we regard an M.V.U.E. as the best possible, and if we face a situation where we cannot establish the existence of such an estimator, there is another possibility open to us. We may be able to establish a lower bound for the variance of an unbiased estimator and compare the variance of some unbiased estimator which is being used with this lower bound. In this way it may be possible to demonstrate that an estimator is nearly as good as possible relative to the criterion of minimum variance unbiasedness. An important result in this connexion is the following theorem.

2.8.1 Theorem (the Cramér–Rao inequality)

Let $\{P_\theta : \theta \in \Theta\}$ be a family of distribution on a sample space X, θ being a real parameter and Θ an interval on the line; and suppose that, for each θ, P_θ is defined by a density function $p(\cdot, \theta)$ with respect to some natural measure on X (see section 1.1.4). Then, subject to certain regularity conditions, the variance of any unbiased estimator $\tilde{\theta}$ of θ satisfies the inequality

$$\operatorname{var}_\theta(\tilde{\theta}) \geq \frac{1}{E_\theta[\{\partial \log p(x, \theta)/\partial \theta\}^2]}.$$

Proof. Since $\tilde{\theta}$ is an unbiased estimator, $E_\theta(\tilde{\theta}) \equiv \theta$,

i.e. $\int_X \tilde{\theta}(x) p(x, \theta) dx = \theta.$

So $\dfrac{\partial}{\partial \theta} \int_X \tilde{\theta}(x) p(x, \theta) dx = 1.$

We now assume sufficient regularity to allow differentiation under the integral sign and so obtain

$$\int_X \tilde{\theta}(x) \frac{\partial p(x, \theta)}{\partial \theta} dx = 1,$$

or $\int_X \tilde{\theta}(x) \dfrac{\partial \log p(x, \theta)}{\partial \theta} p(x, \theta) dx = 1,$

i.e. $E_\theta\left[\tilde{\theta}(x) \dfrac{\partial \log p(x, \theta)}{\partial \theta}\right] = 1.$

Writing $u = \tilde{\theta}(x)$ and $v = \partial \log p(x, \theta)/\partial \theta$, we have then $E_\theta(uv) = 1$.

Now $E_\theta(v) = \int_X \dfrac{\partial \log p(x, \theta)}{\partial \theta} p(x, \theta) dx$

$= \int_X \dfrac{\partial p(x, \theta)}{\partial \theta} dx$

$= \dfrac{\partial}{\partial \theta} \int_X p(x, \theta) dx,$

if again we assume enough regularity to permit differentiation under the integral sign;

and $\int_X p(x, \theta) dx = 1.$

It follows that $E_\theta(v) = 0$

and so $E_\theta(uv) = \operatorname{cov}_\theta(u, v).$

Now $1 = [\operatorname{cov}_\theta(u, v)]^2 \leq \operatorname{var}_\theta(u) \operatorname{var}_\theta(v)$ (see Meyer, 1965, p. 132), and so $\text{var}_\theta(u) \geq \dfrac{1}{\text{var}_\theta(v)}$.

Since $E_\theta(v) = 0$, $\text{var}_\theta(v) = E_\theta(v^2)$ and so, on replacing u and v by $\tilde{\theta}(x)$ and $\partial \log p(\cdot, \theta)/\partial \theta$ respectively, we have

$$\text{var}_\theta(\tilde{\theta}) \geq \dfrac{1}{E_\theta[\{\partial \log p(x,\theta)/\partial \theta\}^2]}.$$

2.8.2 The regularity conditions which we have assumed in deriving this result are concerned with the possibility of differentiating under the integral sign. These can fail for two main reasons: first, the density function $p(x, \theta)$ may not tail off rapidly enough to ensure appropriate convergence of the integral of the differentiated integrand; second, the effective range of integration may depend on θ, the effective range of integration being the set in X for which $p(x, \theta)$ is non-zero. The first reason for insufficient regularity does not give much cause for concern in practice, since realistic density functions do tail off very rapidly. However there are cases of practical importance which will be exemplified later where the theorem fails for the second reason, so that some care must be exercised in its application.

2.9 Fisher's information

The quantity $E_\theta[\{\partial \log p(x,\theta)/\partial \theta\}^2]$ was called by Fisher the *amount of information* about θ contained in an observation in X. It is often denoted by I_θ, and the inequality is then

$$\text{var}_\theta(\tilde{\theta}) \geq I_\theta^{-1}.$$

This demonstrates the reason for the nomenclature. The more information about θ provided on average by an observation, the smaller we might expect the variance of a 'good' estimator to be. It is worth noting also that if $x = (x_1, x_2, \ldots, x_n)$ and the x_is constitute a random sample from a distribution, the information (according to the above definition) provided by the set of x_is is simply n times the information provided by each. For if the distribution from which the sample is drawn has density function $p^*(\cdot, \theta)$ we have

$$\dfrac{\partial}{\partial \theta} \log p(x, \theta) = \sum_{i=1}^{n} \dfrac{\partial}{\partial \theta} \log p^*(x_i, \theta).$$

The right hand side is the sum of n independent, identically distributed, random variables each of which, subject to the kind of regularity demanded in the theorem, has zero mean.

Let $i_\theta = E_\theta\left[\left\{\dfrac{\partial}{\partial \theta} \log p^*(x_i, \theta)\right\}^2\right]$.

This is by definition, the information provided by a single observation. It follows that

$$I_\theta = E_\theta\left[\left\{\frac{\partial}{\partial\theta}\log p(x,\theta)\right\}^2\right] = \text{var}_\theta\left[\frac{\partial}{\partial\theta}\log p(x,\theta)\right]$$

$$= n\,\text{var}_\theta\left[\frac{\partial}{\partial\theta}\log p^*(t,\theta)\right]$$

$$= ni_\theta.$$

2.10 The Cramér–Rao lower bound

Theorem 2.8.1 establishes a lower bound for the variance of an unbiased estimator. It does not tell us how sharp is this lower bound, how nearly it can be attained. The following result shows that only in exceptional circumstances can the bound be attained.

2.10.1 Theorem

Subject to the regularity conditions of Theorem 2.8.1, there exists an unbiased estimator whose variance attains the Cramér–Rao lower bound if and only if

$\partial\log p(x,\theta)/\partial\theta$ *can be expressed in the form* $\partial\log p(x,\theta)/\partial\theta = a(\theta)[\tilde{\theta}(x)-\theta]$ *and in this case* $a(\theta) = I_\theta$.

Proof. In the notation of Theorem 2.8.1, the critical result which led to the lower bound was the fact that

$$[\text{cov}_\theta(u,v)]^2 \leq \text{var}_\theta(u)\,\text{var}_\theta(v).$$

Assuming that neither u nor v is constant, we have equality here if and only if u and v are linearly related (see Meyer, 1965, p. 132); more precisely,

iff $\quad v - E_\theta(v) = a(\theta)[u - E_\theta(u)],$

that is, iff $\quad v = a(\theta)[u-\theta]$

or $\quad \dfrac{\partial}{\partial\theta}\log p(x,\theta) = a(\theta)[\tilde{\theta}(x)-\theta].$

Moreover, in this case,

$$1 = \text{cov}_\theta(u,v) = \frac{\text{var}_\theta(v)}{a(\theta)},$$

so that $\quad a(\theta) = \text{var}_\theta(v) = I_\theta.$

2.10.2 Example

Let $x = (x_1, x_2, \ldots, x_n)$ be the results of n independent trials in each of which the probability of success is θ.

Then $p(x, \theta) = \theta^{m(x)}(1-\theta)^{n-m(x)}$,

where $m(x)$ is the number of successes in the n trials.

$$\frac{\partial}{\partial \theta} \log p(x, \theta) = \frac{m(x)}{\theta} - \frac{n-m(x)}{1-\theta} = \frac{n}{\theta(1-\theta)}[\tilde{\theta}(x)-\theta],$$

where $\tilde{\theta}(x) = \dfrac{m(x)}{n}$ is the proportion of successes in the n trials.

So $m(x)/n$ is an M.V.U.E. of θ and

$$\operatorname{var}_\theta\left(\frac{m(x)}{n}\right) = \frac{\theta(1-\theta)}{n}.$$

This example illustrates the positive application of Theorem 2.10.1 in that it concerns a situation where regularity conditions are easily satisfied and $\partial \log p(x, \theta)/\partial \theta$ can be expressed in the appropriate form; this leads us immediately to an M.V.U.E. However such positive applications of the theorem are few in number. The theorem nevertheless has considerable practical importance, for, as we shall see in a later chapter, in the case of large random samples, $\partial \log p(x, \theta)/\partial \theta$ can often be expressed approximately in the form required by Theorem 2.10.1 and this enables us to find large-sample estimators which are nearly as good as possible, in the M.V.U. sense.

2.11 Efficiency

Even if we are dealing with a situation where Theorem 2.10.1 cannot be used, because $\partial \log p(x, \theta)/\partial \theta$ cannot be expressed in the form there required, it is still possible to use the Cramér–Rao inequality (Theorem 2.8.1) in the following positive way. We may define the *efficiency*, eff $(\tilde{\theta})$, of an unbiased estimator of a real parameter by

$$\operatorname{eff}(\tilde{\theta}) = \frac{I_\theta^{-1}}{\operatorname{var}_\theta(\tilde{\theta})},$$

that is, by comparing its variance with something which we know cannot be bettered. If it turns out that eff $(\tilde{\theta})$ is uniformly (in θ) near 1, this is a positive recommendation in favour of $\tilde{\theta}$ as an estimator. However, we cannot argue that $\tilde{\theta}$ is a poor estimator because eff $(\tilde{\theta})$ is not uniformly near 1, for if we are dealing with a situation in which the lower bound I_θ^{-1} cannot be attained, it is quite possible that the efficiency of every unbiased estimator is low. We have proved nothing to the contrary and indeed the following example demonstrates that this possibility is not just idle speculation.

2.11.1 Example

Consider the case where $x = (x_1, x_2, \ldots, x_n)$ is a random sample from the distribution on the positive numbers with density $\theta^2 x e^{-\theta x}, \theta > 0$. In example 2.6.1 we showed that $(2n-1)/\sum x_i$ is an M.V.U.E. of θ. It is not difficult to calculate that

$$\mathrm{var}_\theta\left[\frac{2n-1}{\sum x_i}\right] = \frac{\theta^2}{2(n-1)},$$

while $I_\theta^{-1} = \dfrac{\theta^2}{2n}.$

If n is large the Cramér–Rao lower bound is nearly attained by the variance of the M.V.U.E. However if n is say 2, the M.V.U.E. has efficiency 0·5 only. So of course every unbiased estimator has low efficiency.

Note. In calculating the information I_θ, the following result is often useful.

2.11.2 Lemma

Subject to regularity conditions,

$$I_\theta = E_\theta\left[-\frac{\partial^2}{\partial \theta^2} \log p(x, \theta)\right].$$

Proof.
$$\frac{\partial^2}{\partial \theta^2} \log p = \frac{\partial}{\partial \theta}\left[\frac{1}{p}\frac{\partial p}{\partial \theta}\right]$$

$$= -\frac{1}{p^2}\left[\frac{\partial p}{\partial \theta}\right]^2 + \frac{1}{p}\frac{\partial^2 p}{\partial \theta^2}$$

$$= -\left[\frac{\partial \log p}{\partial \theta}\right]^2 + \frac{1}{p}\frac{\partial^2 p}{\partial \theta^2}.$$

Now
$$E_\theta\left[\frac{1}{p}\frac{\partial^2 p}{\partial \theta^2}\right] = \int_X \frac{\partial^2 p}{\partial \theta^2} dx$$

$$= \frac{\partial^2}{\partial \theta^2} \int_X p\, dx, \quad \text{given sufficient regularity}$$

$$= \frac{\partial^2}{\partial \theta^2} 1 = 0.$$

Thus $E_\theta\left[-\dfrac{\partial^2}{\partial \theta^2} \log p\right] = E_\theta\left[\left(\dfrac{\partial}{\partial \theta} \log p\right)^2\right] = I_\theta.$

2.12 Generalization of the Cramér–Rao inequality

The Cramér–Rao inequality (Theorem 2.8.1) can be generalized to a vector-valued parameter θ. The analogue, for a parameter $\theta \in \mathbf{R}^s$, of Fisher's information is an information matrix $\mathbf{B}_\theta$, the $s \times s$ matrix whose (i,j)th component is

$$E_\theta \left[\frac{\partial \log p(x, \theta)}{\partial \theta_i} \frac{\partial \log p(x, \theta)}{\partial \theta_j} \right],$$

and subject to regularity conditions as above, we have

$$E_\theta \left[\frac{\partial \log p(x, \theta)}{\partial \theta_i} \frac{\partial \log p(x, \theta)}{\partial \theta_j} \right] = -E_\theta \left[\frac{\partial^2 \log p(x, \theta)}{\partial \theta_i\, \partial \theta_j} \right].$$

The generalization of the Cramér–Rao inequality states that, again subject to regularity conditions, if $\tilde\theta$ is any unbiased estimator of θ, then its variance matrix, the $s \times s$ matrix

$$\mathrm{var}_\theta(\tilde\theta) = E_\theta[(\tilde\theta - \theta)(\tilde\theta - \theta)'],$$

is such that $\mathrm{var}_\theta(\tilde\theta) - \mathbf{B}_\theta^{-1}$ is positive semi-definite. Thus $\mathbf{B}_\theta^{-1}$ is in a sense a 'lower bound' for the variance matrix of an unbiased estimator of $\tilde\theta$.

If we assume sufficient regularity, as in section 2.8.1, the proof of this result is as follows. Let $\tilde\theta = (\tilde\theta_1, \tilde\theta_2, \ldots, \tilde\theta_s)'$ be an unbiased estimator of $\theta = (\theta_1, \theta_2, \ldots, \theta_s)'$.

We have $\int_X \tilde\theta_i(x)\, p(x, \theta)\, dx = \theta_i$,

so that $\int_X \tilde\theta_i(x)\, \dfrac{\partial \log p(x, \theta)}{\partial \theta_j}\, p(x, \theta)\, dx = \delta_{ij}$, the Kronecker delta,

this holding for $i, j = 1, 2, \ldots, s$.

If now we write v for the vector-valued random variable whose ith component is $\partial \log p(x, \theta)/\partial \theta_i$, then:

(i) $E_\theta(\tilde\theta v') = \mathbf{I}$, the unit $s \times s$ matrix; this being the matrix expression of the above equations;
 (ii) $E_\theta(v) = 0$, as in section 2.8.1;
 (iii) $\mathrm{var}_\theta(v) = \mathbf{B}_\theta$, by the definition of $\mathbf{B}_\theta$.

Because of (ii) we may write (i) as $\mathrm{cov}_\theta(\tilde\theta, v) = \mathbf{I}$.

Now consider the variance matrix of the vector $\begin{bmatrix} \tilde\theta \\ v \end{bmatrix}$. This is

$$\begin{bmatrix} \mathrm{var}_\theta(\tilde\theta) & \mathrm{cov}_\theta(\tilde\theta, v) \\ \mathrm{cov}_\theta(v, \tilde\theta) & \mathrm{var}_\theta(v) \end{bmatrix} = \begin{bmatrix} \mathrm{var}_\theta(\tilde\theta) & \mathbf{I} \\ \mathbf{I} & \mathbf{B}_\theta \end{bmatrix}.$$

Since the matrix on the right side is a variance matrix, it is positive semi-definite. It follows that the matrix

$$\begin{bmatrix} I & -B_\theta^{-1} \end{bmatrix} \begin{bmatrix} \mathrm{var}_\theta(\tilde{\theta}) & I \\ I & B_\theta \end{bmatrix} \begin{bmatrix} I \\ -B_\theta^{-1} \end{bmatrix}$$

is also positive semi-definite, and, as is easily verified, this last matrix is simply $\mathrm{var}_\theta(\tilde{\theta}) - B_\theta^{-1}$.

2.12.1 Another way of looking at this generalization is worth noting. Let

$$c'\theta = c_1\theta_1 + c_2\theta_2 + \ldots + c_s\theta_s$$

be a linear combination of the unknown components of the vector θ, the c_is being known. Then if $\tilde{\theta}$ is an unbiased estimator of θ, $c'\tilde{\theta}$ is an unbiased estimator of $c'\theta$ and

$$\mathrm{var}_\theta(c'\tilde{\theta}) = c' \, \mathrm{var}_\theta(\tilde{\theta}) \, c.$$

The above result states that

$$\mathrm{var}_\theta(c'\tilde{\theta}) \geqslant c' B_\theta^{-1} c.$$

Hence if we can find an unbiased estimator $\tilde{\theta}$ such that $\mathrm{var}_\theta(\tilde{\theta}) = B_\theta^{-1}$ this estimator has in this linear sense smaller dispersion than any other unbiased estimator of θ.

2.12.2 The following is a very simple illustration of the above results, which does not really exhibit their usefulness in practice, but which avoids the complicated manipulation involved in a more realistic application. The reader may use it to verify general statements involved in the above proofs.

A random sample of n individuals is drawn from a very large population in which each individual falls into one of three classes. Of the n drawn, x_i are in class i, $i = 1, 2, 3$, so that $x_1 + x_2 + x_3 = n$. The population proportions in the three classes are θ_1, θ_2 and $1 - \theta_1 - \theta_2$, respectively. Determine a 'lower bound' for the variance matrix of unbiased estimators of θ_1 and θ_2.

Here $x = (x_1, x_2, x_3)$, $\theta = (\theta_1, \theta_2)$,

$$p(x, \theta) = \frac{n!}{x_1! \, x_2! \, x_3!} \theta_1^{x_1} \theta_2^{x_2} (1 - \theta_1 - \theta_2)^{x_3}.$$

It is easily verified that

$$\frac{\partial^2 \log p(x, \theta)}{\partial \theta_1^2} = -\frac{x_1}{\theta_1^2} - \frac{x_3}{(1 - \theta_1 - \theta_2)^2}.$$

Now $E_\theta(x_1) = n\theta_1$ and $E_\theta(x_3) = n(1 - \theta_1 - \theta_2)$,

so that $\displaystyle E_\theta\left[-\frac{\partial^2 \log p(x, \theta)}{\partial \theta_1^2}\right] = n\left[\frac{1}{\theta_1} + \frac{1}{1 - \theta_1 - \theta_2}\right].$

From similar calculations we find that

$$\mathbf{B}_\theta = n \begin{bmatrix} \dfrac{1}{\theta_1} + \dfrac{1}{1-\theta_1-\theta_2} & \dfrac{1}{1-\theta_1-\theta_2} \\ \dfrac{1}{1-\theta_1-\theta_2} & \dfrac{1}{\theta_2} + \dfrac{1}{1-\theta_1-\theta_2} \end{bmatrix}.$$

It is not difficult to show that

$$\mathbf{B}_\theta^{-1} = \frac{1}{n} \begin{bmatrix} \theta_1(1-\theta_1) & -\theta_1\theta_2 \\ -\theta_1\theta_2 & \theta_2(1-\theta_2) \end{bmatrix}.$$

and this is the required lower bound.

It may be verified that in this case the lower bound is attained by the variance matrix of the 'obvious' unbiased estimators

$$\hat{\theta}_1(x) = \frac{x_1}{n}, \qquad \hat{\theta}_2(x) = \frac{x_2}{n}.$$

2.13 Concluding remarks

In this chapter we have adopted the point of view that a good estimator of a real parameter is one which is minimum-variance unbiased. It must be re-emphasized that there is an element of arbitrariness in this criterion, particularly with regard to unbiasedness, which was forced on us by the consideration that, in general, estimators of uniformly minimum mean-square error do not exist. However, granted in the meantime that this is a generally accepted criterion, how far have we progressed in the problem of establishing the existence of, and calculating, M.V.U.E.s? The answer is, 'Not very far.' There are many situations where either no M.V.U.E. exists or where we cannot establish whether or not such an estimator exists. A simple illustration of the former case is the fact that, if θ is the probability of success in a trial, there exists no unbiased estimator of the odds in favour of success, namely $\theta/(1-\theta)$, based on the results of n independent trials (see Hodges and Lehmann, 1964, p. 217). Also it is generally true that if the family of distributions of a minimal-sufficient statistic is not complete (see Example 2.7.1), then it is extremely difficult to establish whether or not an M.V.U.E. exists.

It would be possible to pursue this line of thinking further, to investigate other conditions which might ensure the existence of an M.V.U.E., or to adopt a new criterion which might cover situations where no M.V.U.E. exists and to investigate its implications; but statistics is primarily an applied subject. Observers wish their results analysed and inferences made, and so methods have been developed which have a strong intuitive appeal, but which cannot always be justified in the kind of terms which we have been discussing. We shall now turn to two important methods of estimation and we shall investigate to what extent these methods can be justified by the criterion of minimum-variance unbiasedness.

Examples

2.1 Let $x_1, x_2, \ldots, x_n$ be a random sample from a Poisson distribution with unknown mean θ. Find an M.V.U.E. of $e^{-\theta}$, the probability of the zero class. Determine also its variance.

2.2 Let $x_1, x_2, \ldots, x_n$, where $n > 3$, be a random sample from an $N(\mu, \sigma^2)$ distribution with μ and σ^2 unknown. Find an M.V.U.E. of μ^2/σ^2.

2.3 If $x_1, x_2, \ldots, x_n$ are independent random variables each with probability density $\theta e^{-x\theta}$ ($x > 0$, $\theta > 0$), show that the random variable which takes the value 1 when $x_1 \geq k$ and the value 0 when $x_1 < k$, is an unbiased estimator of $g(\theta) = e^{-k\theta}$. Hence show that, for a suitable choice of statistic t,

$$\hat{g}(t) = \begin{cases} 0 & \text{when } t < k \\ \left[\dfrac{t-k}{t}\right]^{n-1} & \text{when } t \geq k \end{cases}$$

is an M.V.U.E. of g. (*Camb. Dip.*)

2.4 If $x_1, x_2, \ldots, x_n$ is a random sample from an $N(\mu, \sigma^2)$ distribution, show that the estimator

$$\frac{1}{n}\sum_{i=1}^{n}(x_i - \bar{x})^2$$

of σ^2 has smaller mean-square error than that of the M.V.U.E.,

$$\frac{1}{n-1}\sum_{i=1}^{n}(x_i - \bar{x})^2.$$

2.5 Let $x_1, x_2, \ldots, x_n$ be a random sample from the uniform distribution on the interval $(0, \theta)$. Show that $\max(x_1, x_2, \ldots, x_n)$ is sufficient for θ and that an unbiased estimator for θ of the form $k \max(x_1, x_2, \ldots, x_n)$ exists. Determine this unbiased estimator and find its variance. Compare this variance with the Cramér–Rao lower bound for the variance of an unbiased estimator and explain why this lower bound is not applicable in this instance.

2.6 A random sample of size n is available from the distribution on the positive real numbers with density

$$\frac{x+1}{\theta(\theta+1)} e^{-x/\theta},$$

where θ is an unknown positive parameter. Obtain an unbiased estimator of $(3+2\theta)(2+\theta)/(\theta+1)$ whose variance attains the Cramér–Rao lower bound. (*Camb. Dip.*)

2.7 An entomologist samples at random from a large population of a particular species. He records the sex of each insect and stops sampling when he has obtained M males ($M > 1$), by which time his total sample size is x. Find an M.V.U.E. of θ, the proportion of males in the population. (*Camb. Dip.*)

2.8 Leaves of a plant are examined for insects and it is found that x_i leaves have precisely i insects ($i = 1, 2, \ldots$; $\sum x_i = N$). The number of insects per leaf is believed to be Poisson, except that many leaves have no insects because they are unsuitable for feeding and not merely because of the chance variation allowed for by the Poisson distribution. The empty leaves are therefore not counted. Show that

$$\sum_{i=2}^{\infty} \frac{ix_i}{N}$$

is an unbiased estimator of the Poisson parameter μ, and determine its efficiency. (*Camb. Dip.*)

3 The Method of Least Squares

3.1 Examples

The intuitive appeal of the method of least squares may be illustrated by two simple examples.

3.1.1 Suppose that $x_1, x_2, \ldots, x_n$ is a random sample from a distribution on the line and that we are interested in estimating the mean, θ, of this distribution. We may write

$$x_i = \theta + \varepsilon_i,$$

where ε_i is the deviation or error of the observation x_i from the mean of the underlying distribution. A reasonably optimistic attitude here is to assume that the observations are trying to give us information about θ and that the deviations $\varepsilon_1, \varepsilon_2, \ldots, \varepsilon_n$ are in some sense small. Therefore a plausible method of estimating θ is to choose as estimate a number for which these deviations are small, and one way of doing so is to choose as estimate a value of θ for which $\sum_{i=1}^{n} \varepsilon_i^2$ is as small as possible: that is, to estimate θ by the number $\hat{\theta}(x_1, x_2, \ldots, x_n)$ which minimizes $\sum_{i=1}^{n} (x_i - \theta)^2$, regarded as a function of θ. If we do this, then of course

$$\hat{\theta}(x_1, x_2, \ldots, x_n) = \frac{1}{n}(x_1 + x_2 + \ldots + x_n) = \bar{x},$$

the mean of the observations, so that our plausible method leads to an intuitively appealing result in this case.

3.1.2 As another example, suppose that observations $x_1, x_2, \ldots, x_n$ are made at different values $a_1, a_2, \ldots, a_n$, respectively, of a 'concomitant variable' a. The xs are random variables and the as are known, non-random numbers. For instance the a_is might be different levels of a fertilizer and the x_is corresponding yields of a crop. The as are then controlled by the observer but there are factors affecting the xs outwith the observer's control – weather for instance. Suppose we know that 'the mean of x varies linearly with a', but we do not know the exact form of this relationship. In other words we know that

$E(x_i) = \alpha + \beta a_i$, though α and β are unknown. Then we may write

$$x_i = \alpha + \beta a_i + \varepsilon_i,$$

and, acting on the same optimistic principle as previously, estimate α and β by these numbers which minimize

$$\sum_{i=1}^n \varepsilon_i^2 = \sum_{i=1}^n (x_i - \alpha - \beta a_i)^2.$$

This method is again appealing provided that the deviations $\varepsilon_1, \varepsilon_2, \ldots, \varepsilon_n$ all have 'the same chance of being small', but if some have more chance of being small than others it might seem more sensible to estimate α and β by minimizing some weighted sum of squares

$$\sum_{i=1}^n w_i (x_i - \alpha - \beta a_i)^2,$$

the ws being weights which are large for those is for which ε_i is liable to be small and small for ε_is liable to be large. Another complicating factor which might lead us to think again about this method of estimation is the possibility of interdependencies among the ε_is. It is not then so obvious how we might adjust the method. However a mathematical investigation of the properties of this method will suggest an appropriate adjustment.

3.2 Normal equations

These two examples are particular cases of the following general situation. A random vector $x = (x_1, x_2, \ldots, x_n)'$ is such that

$$x = A\beta + \varepsilon,$$

where A is a known matrix of order $n \times p$ with $p < n$, $\beta = (\beta_1, \beta_2, \ldots, \beta_p)'$ is a vector of unknown parameters and $\varepsilon = (\varepsilon_1, \varepsilon_2, \ldots, \varepsilon_n)'$ is a vector of 'deviations from the mean' or 'errors', that is, a vector whose expected value is zero. In this general situation we may apply the principle used in the examples above and estimate β by minimizing the sum of squares

$$\sum_{i=1}^n \varepsilon_i^2 = \varepsilon'\varepsilon = (x - A\beta)'(x - A\beta).$$

This method of estimation is called the *method of least squares*, for obvious reasons, and any minimizing value $\hat\beta(x)$ is called a *least-square estimate* of β. The function $\hat\beta$, a function from R^n (Euclidean n-space) into R^p, is a least-squares estimator. However we shall not maintain the notational distinction between $\hat\beta(x)$ and $\hat\beta$ and we shall use the latter for both. The context will make clear the sense in which it is used.

Determination of a least-squares estimate is not a difficult problem. We have to choose $\hat{\beta}$ to minimize the quadratic form

$$(x - A\beta)'(x - A\beta)$$

in the components $\beta_1, \beta_2, \ldots, \beta_p$ of β. Differentiation of this quadratic form with respect to $\beta_1, \beta_2, \ldots, \beta_p$ leads to the so-called *normal equations* satisfied by a least-squares estimate, namely

$$A'A\beta = A'x,$$

and any solution of these equations does in fact minimize $(x - A\beta)'(x - A\beta)$ and so is a least-squares estimate (see Appendix A). If rank $A = p$, then $A'A$, which has the same rank as A (see Appendix A), is non-singular and there is a unique least-squares estimate

$$\hat{\beta} = (A'A)^{-1}A'x.$$

If rank $A < p$, then $A'A$ is singular, the normal equations do not have a unique solution, and there is a family of least-squares estimates which may be determined in any particular case by the usual methods for solving a system of linear equations.

3.3 Geometric interpretation

The intuitive appreciation of linear algebra is greatly aided by a geometrical interpretation in which vectors are represented by points and matrices are regarded as representations of linear transformations or functions (Hohn, 1964, p. 182). This applies equally to the *linear statistical model*, $x = A\beta + \varepsilon$, which we are investigating.

The sample space here is R^n and there is a true distribution on this space which we do not know. We do have some knowledge about the mean vector or *centre* θ of this distribution, for we know that $\theta = E(x)$ can be expressed in the form $A\beta$; in other words that θ lies in a subspace ω of R^n, the subspace spanned by the columns of A, which we shall refer to as the range of A. Given an observation x, we estimate θ by $\hat{\theta} = A\hat{\beta}$.

Now $x - \hat{\theta}$ is orthogonal to ω, since $A'(x - \hat{\theta}) = A'x - A'A\hat{\beta} = 0$, so that $x - \hat{\theta}$ is orthogonal to every vector of the form $A\beta$. This means that we estimate θ by the projection of x on ω, or by the point of ω nearest to x; and this seems reasonable on the grounds that x is probably near the centre of the distribution. Here we have the geometric essence of the method of least squares.

Of course $\hat{\theta}$, the projection of x on ω, is always unique whatever the rank of the matrix A. On the other hand, any point of R^p which is mapped by A into $\hat{\theta}$ is a least-squares estimate of β. If A has rank p it establishes a one-to-one correspondence between points in R^p and points in ω and then $\hat{\beta}$ is unique. In this case $\hat{\theta} = A(A'A)^{-1}A'x$. The matrix $A(A'A)^{-1}A'$ is symmetric and idempotent and it represents the orthogonal projection of R^n on to the range ω of A (see Appendix A). If rank $A < p$, then A establishes a many-to-one correspondence between points β of $\mathbf{R}^p$ and points θ of ω, whose dimension is rank A, so that while $\hat{\theta}$ is still unique, $\hat{\beta}$ is not. There is not then a simple matrix representation for the projection of $\mathbf{R}^n$ on ω.

For the case $n = 3, p = 2$, we can actually draw pictures for this geometric interpretation.

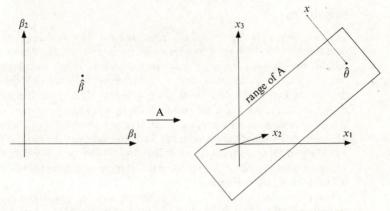

Figure 1 The matrix A of order 3×2 and rank 2 maps $\mathbf{R}^2$ onto a two-dimensional subspace of $\mathbf{R}^3$

Figure 1 gives a geometric picture of the case where the observation vector x has dimension 3, β has dimension 2 and the matrix A (which then has order 3×2) is of rank 2. A may be regarded as representing a linear transformation from $\mathbf{R}^2$ into $\mathbf{R}^3$, and, because its rank is 2, its range is represented by a plane in $\mathbf{R}^3$ (see Hohn, 1964, p. 182). $\hat{\theta}$ is the projection of x on this plane, and there is a unique $\hat{\beta}$ mapped by A onto $\hat{\theta}$.

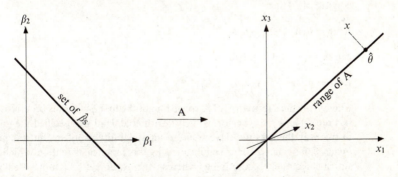

Figure 2 The matrix A of order 3×2 and rank 1 maps $\mathbf{R}^2$ onto a one-dimensional subspace of $\mathbf{R}^3$

Figure 2 illustrates the same case except that now, A has rank 1. Its range then is represented by a line through the origin. $\hat{\theta}$ is the projection of x on this line. But now there is not a unique $\hat{\beta}$ mapped by A onto $\hat{\theta}$, and the set of $\hat{\beta}$ such that $A\hat{\beta} = \hat{\theta}$ is represented by a line in $\mathbf{R}^2$ as indicated.

3.4 Identifiability

The case where rank A $< p$ in the linear model $x = A\beta + \varepsilon$ raises for the first time an issue which will be of concern later. If we are given a distribution for ε, the distribution on the sample space depends on β, as this distribution is centred on $A\beta$. However when rank A $< p$, different values of β yield the same distribution on the sample space because different values of β correspond to the same value of $A\beta$. It is clear that in this case, while an observation x may give us some information about $A\beta$, it can give no discriminatory information whatsoever between different values of β corresponding to the same value of $A\beta$. The parameter β is said to be unidentifiable. More generally, if different values of some parameter give the same distribution on the sample space, this parameter is not identifiable.

When a parameter is not identifiable we may say that two values of it are equivalent if and only if they yield the same distribution on the sample space. This defines an equivalence relation which partitions the parameter space into equivalence classes. Usually then an observation will give information about which equivalence class the true parameter belongs to but no information about which member of this equivalence class the true parameter is. This difficulty really arises because of our specification of the statistical model describing the situation in which observations are made, and it can be avoided by a different specification of the model. For instance, in the above linear model if rank A $= q < p$, then $p-q$ of the columns of A, say the last $p-q$, are linear combinations of the remaining q. It follows that, if $\mathbf{a}_i$ is the ith column of A, then

$$E(x) = \beta_1 \mathbf{a}_1 + \ldots + \beta_p \mathbf{a}_p$$

can be expressed in the form

$$E(x) = \gamma_1 \mathbf{a}_1 + \ldots + \gamma_q \mathbf{a}_q = A_q \gamma,$$

where A_q is the sub-matrix of A consisting of the first q columns of A. Now A_q has rank q and γ, a q-vector, is then identifiable. Had we specified the model in this way there would have been no identifiability problem; but the parameter β may have some significance in the practical situation which the statistical model is describing, whereas the parameter γ is not so easy to interpret. 'Natural' parametrization in the model set up may lead to non-identifiability, which is more in the nature of an irritant than a source of deep problems.

3.5 The Gauss–Markov theorem

The method of least squares has been introduced on the purely intuitive basis that if we estimate the mean of a distribution by the parameter nearest the observation made, this estimate should be quite good, since the observation is probably near the true mean. Some stronger justification of the method is really desirable and such a justification is provided by the celebrated theorem which we shall discuss in this section.

Whereas in chapter 2 we were dealing with real-valued parameters and estimates of these, we are concerned here with a vector-valued parameter β. Our criterion for choosing among unbiased estimates of a real parameter was that of minimum variance; in other words if $\hat{\theta}$ and $\tilde{\theta}$ were two unbiased estimates, $\hat{\theta}$ was regarded as better than $\tilde{\theta}$ if $\mathrm{var}_\theta(\tilde{\theta}) - \mathrm{var}_\theta(\hat{\theta})$ were greater than 0. What is the analogue of this criterion for a vector-valued parameter? It is that if $\hat{\beta}$ and $\tilde{\beta}$ are unbiased estimates of the vector β and their *variance matrices* are $\mathrm{var}_\beta(\hat{\beta}) = E_\beta(\hat{\beta} - \beta)(\hat{\beta} - \beta)'$ and $\mathrm{var}_\beta(\tilde{\beta})$, similarly defined, then $\hat{\beta}$ is a better estimate than $\tilde{\beta}$ if the *matrix* $\mathrm{var}_\beta(\tilde{\beta}) - \mathrm{var}_\beta(\hat{\beta})$ is positive semi-definite for all β. The dispersion of the random vector $\hat{\beta}$ about its mean β is then smaller than that of $\tilde{\beta}$. Another way of putting this is to say that the variance of any linear combination of the components of $\hat{\beta}$ is no larger than that of the same linear combination of the components of $\tilde{\beta}$; symbolically that, for every p-vector c,

$$\mathrm{var}_\beta(c'\hat{\beta}) \leq \mathrm{var}_\beta(c'\tilde{\beta}).$$

Now, in the identifiable case at least, the least-squares estimate $\hat{\beta}$ of β is a linear estimate in the sense that $\hat{\beta}_i(x)$ is a linear combination of the components of the observation x. It is also unbiased, since

$$E_\beta(\hat{\beta}) = E_\beta\{(A'A)^{-1}A'x\} = (A'A)^{-1}A'E_\beta(x) = (A'A)^{-1}A'A\beta = \beta.$$

The Gauss–Markov theorem proves that subject to certain conditions on the error-vector ε the least squares estimate $\hat{\beta}$ is better in the above sense than any other unbiased *linear* estimate. This of course is a weaker result than one which states that an estimator is best in the class of *all* unbiased estimates, but the input in the way of assumptions concerning the error vector is weak as we shall see, and as a general rule in the theory of inference, the weaker the assumptions, or the wider the family of possible distributions is allowed to be, the weaker are the results which can be obtained.

The examples in section 3.1 suggest that in order that the least-squares estimate be best, it may be necessary to require that the components of the error vector be independent and identically distributed. In fact the Gauss–Markov theorem requires less than this – only that these components have the same variance and are uncorrelated. Since the possibility of non-identifiability of β complicates matters considerably, we give first a proof for the case where rank $A = p$ so that β is identifiable and the least-squares estimate is unique and equal to $(A'A)^{-1}A'x$.

3.5.1 The case where β is identifiable

Let x be a random n-vector expressible in the form $x = A\beta + \varepsilon$ where A is a known $n \times p$ matrix of rank p, β is an unknown p-vector and ε is an error vector with $E(\varepsilon) = 0$ and $\text{var}(\varepsilon) = \sigma^2 I$, where σ^2 is unknown, that is the components of ε have the same unknown variance σ^2 and are uncorrelated. Let $\hat{\beta}$ be the unique least-squares estimator of β and let $\phi = c'\beta$ be a linear parametric function. Then $c'\hat{\beta}$ is an unbiased estimator of ϕ and, if $\tilde{\phi}$ is any other linear unbiased estimator of ϕ, we have $\text{var}_\beta(c'\hat{\beta}) \leqslant \text{var}_\beta(\tilde{\phi})$.

Proof. $E_\beta(c'\hat{\beta}) = c'E_\beta(\hat{\beta}) = c'\beta = \phi$ so that $c'\hat{\beta}$ is unbiased.

Since $\tilde{\phi}$ is a linear estimator it is expressible in the form $b'x$ and since $\tilde{\phi}$ is an unbiased estimator of ϕ, we have

$$b'A\beta = b'\{E_\beta(x)\} = E_\beta(b'x) = E_\beta(\phi) = c'\beta \quad \text{for every } \beta.$$

Hence $b'A = c'$.

Now $\text{var}_\beta(\tilde{\phi}) = \text{var}_\beta(b'x) = b'(\text{var}_\beta x)b = \sigma^2 b'b$,

and similarly

$$\text{var}_\beta(\hat{\beta}) = \text{var}_\beta\{(A'A)^{-1}A'x\} = (A'A)^{-1}A' \text{var}_\beta(x) A(A'A)^{-1} = \sigma^2(A'A)^{-1}.$$

It follows that $\text{var}_\beta(c'\hat{\beta}) = \sigma^2 c'(A'A)^{-1}c = \sigma^2 b'A(A'A)^{-1}A'b$.

To prove the theorem we must therefore show that

$$b'b \geqslant b'A(A'A)^{-1}A'b,$$

or that $I - A(A'A)^{-1}A'$ is positive semi-definite. This follows from the easily verifiable fact that this matrix is idempotent. (Incidentally it represents the orthogonal projection of R^n on to the orthogonal complement of the range of A and $b'\{I - A(A'A)^{-1}A'\}b$ is the square of the distance of the vector b from the range of A.)

This completes the proof.

3.5.2 The general case

If rank $A < p$, two complications arise in the above proof. The more obvious of these is that $A'A$ is then singular and we do not have the previous simple expression for a least-squares estimate $\hat{\beta}$. It remains true however that any least-squares estimate satisfies the equation

$A'A\hat{\beta} = A'x.$

We note, for subsequent use, that this implies that if a is a vector in the range of A, that is, a vector which can be expressed as a linear combination of the columns of A,

then $\quad a'A\hat{\beta} = a'x.$

The second and less obvious complication is the fact that when rank $A < p$, not every linear parametric function possesses an unbiased linear estimator, for in order that $b'x$ should be an unbiased estimator of $c'\beta$ we require (see Theorem 3.5.1) that $c' = b'A$ or that c' should be expressible as a linear combination of the rows of A. When rank $A = p$, the rows of A span R^p and every p-vector c' can be expressed as a linear combination of these rows. This is not so when rank $A < p$. In considering the general case therefore, we must restrict attention to linear parametric functions $c'\beta$ which have unbiased linear estimators or are *estimable*.

Suppose then that $\phi = c'\beta$ is an estimable parametric function. We wish to show that, if $\hat{\beta}$ is any least-squares estimate of β, $c'\hat{\beta}$ is an unbiased estimator of ϕ and that

$$\text{var}_\beta(c'\hat{\beta}) < \text{var}^\beta(\tilde{\phi})$$

for any other unbiased linear estimator $\tilde{\phi}$ of ϕ; this being subject to the conditions of Theorem 3.5.1 apart from that on rank A. Now since ϕ is estimable there exists a $b \in R^n$ such that $c' = b'A$. Let a be the projection of b on the range of A so that $b-a$ is orthogonal to the range of A, or $(b-a)'A = 0$. Then $a'x$ also is an unbiased estimator of ϕ since

$$E_\beta(a'x) = E_\beta\{(a-b)'x+b'x\} = (a-b)'A\beta+\phi = \phi.$$

Moreover a is the *only* vector in range A such that $a'x$ is an unbiased estimator of ϕ. For suppose there is another such vector a^*. Then we have for every β,

$$E_\beta\{(a-a^*)'x\} = 0$$
i.e. $(a-a^*)'A\beta = 0$
and so $(a-a^*)'A = 0$.

This means that $a-a^*$ is orthogonal to range A, but since a and a^* are both in range A, so is $a-a^*$. Therefore $a-a^* = 0$, i.e. $a = a^*$.

Now $\text{var}_\beta(a'x) \leq \text{var}_\beta(b'x)$, where $b'x$ is any other unbiased estimator of ϕ, since $\text{var}_\beta(a'x) = \sigma^2 a'a$,
while $\text{var}_\beta(b'x) = \sigma^2 b'b = \sigma^2\{a'a+(b-a)'(b-a)\} \geq \sigma^2 a'a$.

We now complete the proof of the fact that

$$\text{var}_\beta(c'\hat{\beta}) \leq \text{var}_\beta(\tilde{\phi})$$

by showing that $\text{var}_\beta(c'\hat{\beta}) = \text{var}_\beta(a'x)$.

Since $a'x$ is an unbiased estimator of ϕ, we have $a'A = c'$.

Therefore $c'\hat{\beta} = a'A\hat{\beta} = a'x$,

since a is in range A and $\hat{\beta}$ satisfies the normal equations. It follows that

$$\text{var}_\beta(c'\hat{\beta}) = \sigma^2 a'a.$$

It remains true in the general case, that any least-squares estimate is better than any other unbiased linear estimator in this special sense that it leads to unbiased estimators of estimable linear functions with smaller variance.

3.5.3 *Remarks*

There are many statements and proofs of this celebrated theorem with various degrees of generality. The proof given in section 3.5.2, which incidentally is valid also when rank A = p, is essentially the same as that given by Scheffé (1960).

It will be seen that in the general proof (contrary to the proof of section 3.5.1) we have never explicitly stated that any least-squares estimator $\hat{\beta}$ of β is unbiased. The reason for this is as follows. We may regard any component of β, say its first component β_1, as a linear parametric function – the function $c'\beta$ where $c' = (1, 0, 0, \ldots, 0)$. It may be that this is not estimable. In the unlikely event that the first column of A consists entirely of zeros, it is immediately obvious that no observation gives any information about β_1 and that β_1 is not estimable. So in this case, as there exists no unbiased estimator of β_1, *a fortiori* there exists none of β. However it is true that if a component β_i of β is estimable, the corresponding component of any least-squares estimate of β is an unbiased estimator of β_i. – this follows from the fact established in section 3.5.2 that $c'\hat{\beta} = a'x$, in the notation of that section.

3.6 Weighted least squares

In the last paragraph of section 3.1 we anticipated the possibility that least-squares estimators might not be 'best' when the components of the error vector ε did not all 'have the same chance of being small', and indeed a crucial part is played in the above proofs by the assumption that the variances matrix of the error vector is $\sigma^2 I$, that is, that its components have the same variance and are uncorrelated. The algebra of the Gauss–Markov theorem suggests the appropriate modification to the method of least squares when either the errors have different variances or when they are correlated.

Suppose then that we consider the linear model

$$x = A\beta + \varepsilon$$

with the same assumptions as before except that now, instead of having var $\varepsilon = \sigma^2 I$, we assume that var $\varepsilon = \sigma^2 \Sigma$, where Σ is a *known* positive definite matrix. This allows for the possibility of differing variances among the ε_is and for correlation between them. By a non-singular linear transformation we can transform this model to that previously investigated by the Gauss–Markov theorem. For since Σ is positive definite, it can be expressed in the form PP' where P is non-singular. Now let $\eta = P^{-1}\varepsilon$ and $y = P^{-1}x$. Then we may write the model in the form

$$Py = A\beta + P\eta$$
or $\quad y = P^{-1}A\beta + \eta = B\beta + \eta, \quad$ say.

Also var $\eta = P^{-1}(\text{var } \varepsilon)P'^{-1} = \sigma^2 P^{-1} PP'P'^{-1} = \sigma^2 I$, so that in terms of y, B and η the model is just that already discussed. We therefore obtain a 'best' estimator of β (that is, an unbiased estimator with minimum dispersion, in the sense of the Gauss–Markov theorem, among the class of linear unbiased estimators) by minimizing the sum of squares

$(y - B\beta)'(y - B\beta)$.

Now $(y - B\beta)'(y - B\beta) = (x - A\beta)(PP')^{-1}(x - A\beta)$
$= (x - A\beta)\Sigma^{-1}(x - A\beta)$.

Hence in the case where the error components have variance matrix $\sigma^2\Sigma$, we obtain best estimators by minimizing this quadratic form rather than the straight sum of squares. In particular, suppose that the components of the error vector are uncorrelated, but have unequal variances so that

$\Sigma = \text{diag}\{\sigma_1^2, \sigma_2^2, \ldots, \sigma_n^2\}$, say.

The expression which we minimize may be written

$$\sum_{i=1}^{n} \frac{(x_i - a_{i1}\beta_1 - a_{i2}\beta_2 - \ldots - a_{ip}\beta_p)^2}{\sigma_i^2}.$$

In other words we weight each square in the sum by the inverse of the variance of the corresponding error component. In this way, as anticipated, we give more weight to the errors which are liable to be small.

It is worth remarking that even when var ε cannot be expressed in the form $\sigma^2 I$, but has the more general form $\sigma^2\Sigma$, an estimator obtained by minimizing the straight sum of squares $(x - A\beta)'(x - A\beta)$ is still unbiased. For instance, in the case where A has rank p, this estimator $\hat{\beta}$ is given by

$\hat{\beta} = (A'A)^{-1} A'x$

and since $E_\beta(x) = A\beta$ we still have $E_\beta(\hat{\beta}) = \beta$. However, in this case

var $\hat{\beta} = \sigma^2 (A'A)^{-1}(A'\Sigma A)(A'A)^{-1}$,

whereas if $\tilde{\beta}$ is the estimator minimizing

$(x - A\beta)'\Sigma^{-1}(x - A\beta) = (y - B\beta)'(y - B\beta)$,

we have

var $\tilde{\beta} = \sigma^2 (B'B)^{-1} = \sigma^2 (A'\Sigma^{-1}A)^{-1}$.

The Gauss–Markov theorem tells us that the matrix

$(A'A)^{-1}(A'\Sigma A)(A'A)^{-1} - (A'\Sigma^{-1}A)^{-1}$

is positive semi-definite, so that, in particular, the diagonal elements of this matrix are all non-negative. This means that the variance of any component of $\hat{\beta}$ is at least as large as that of the corresponding component of $\tilde{\beta}$, and it may well be considerably larger. So while straight least squares yields unbiased estimators of the components of β in this situation, it may be very inefficient in the sense that the variances of these estimators are unnecessarily large, relative to the best we can achieve using linear estimators.

3.7 Estimation of σ^2

In practice when the model

$$x = A\beta + \varepsilon$$

with var $\varepsilon = \sigma^2 I_n$, is appropriate for describing some observational situation, not only will β be unknown, but so also will σ^2. We are then faced with the problem of estimating this unknown quantity as well as β. If $\hat{\beta}$ is any least-squares estimate of β then we might expect that the *residual sum of squares*

$$(x - A\hat{\beta})'(x - A\hat{\beta})$$

will on average increase and decrease with σ^2. And indeed it is not difficult to construct an unbiased estimator of σ^2 from this residual sum of squares. We proceed as follows.

We have $\varepsilon = (x - A\hat{\beta}) + A(\hat{\beta} - \beta)$.

Now $A(\hat{\beta} - \beta)$ is a vector in range A, while $(x - A\hat{\beta})$ is orthogonal to range A (since $A'(x - A\hat{\beta}) = 0$). It follows that if we change the basis in R^n to a new orthonormal basis whose first r elements are in range A (r = rank A) and whose remaining $n - r$ elements are orthogonal to range A and if, under this transformation,

$$(\varepsilon_1, \varepsilon_2, \ldots, \varepsilon_n) \to (\eta_1, \eta_2, \ldots, \eta_n),$$

then $A(\hat{\beta} - \beta) \to (\eta_1, \eta_2, \ldots, \eta_r, 0, 0, \ldots, 0)'$

and $(x - A\hat{\beta}) \to (0, 0, \ldots, 0, \eta_{r+1}, \eta_{r+2}, \ldots, \eta_n)'$.

Hence $(x - A\hat{\beta})'(x - A\hat{\beta}) = \eta_{r+1}^2 + \eta_{r+2}^2 + \ldots + \eta_n^2$.

If uncorrelated random variables with zero means and common variance σ^2 are subjected to an orthogonal transformation, the resulting random variables have the same properties. ($\eta = P\varepsilon$, where $PP' = I_n$; $E(\eta) = PE(\varepsilon) = 0$ and var $\eta = P$ var $\varepsilon P' = \sigma^2 PP' = \sigma^2 I_n$.)

So $E(\eta_i^2) = \sigma^2$, for all i.

Hence $E_\beta\{(x - A\hat{\beta})'(x - A\hat{\beta})\} = E\left(\sum_{i=r+1}^{n} \eta_i^2\right) = (n - r)\sigma^2$,

so that $\dfrac{1}{n-r}(x - A\hat{\beta})'(x - A\hat{\beta})$

is an unbiased estimator of σ^2.

The modification required when var $\varepsilon = \sigma^2 \Sigma$, with Σ known, instead of $\sigma^2 I_n$, is clear. In the notation of section 3.6,

$$\frac{1}{n-r}(y - B\tilde{\beta})'(y - B\tilde{\beta}) = \frac{1}{n-r}(x - A\tilde{\beta})' \Sigma^{-1}(x - A\tilde{\beta})$$

is then the corresponding unbiased estimator of σ^2.

3.8 Variance of least-squares estimators

Another practical consideration which we must take into account is as follows. There is little point in practice in knowing that an estimator of an unknown parameter is best in some sense without the additional knowledge of how near to the parameter our estimate is liable to be. We shall consider this question in general later and in the meantime we content ourselves with the remark that the variance of an estimator gives some idea of its reliability or accuracy. Thus when an estimate is given in practice it is usual to quote also its standard deviation or an estimate of its standard deviation.

Suppose then that we consider the linear model

$$x = A\beta + \varepsilon,$$

where var $\varepsilon = \sigma^2 I_n$ and A has full rank ($= p$) so that no identifiability problems arise. Then the unique least-squares estimate $\hat{\beta}$ of β is given by

$$\hat{\beta} = (A'A)^{-1} A'x.$$

We can deduce immediately that

$$\text{var } \hat{\beta} = (A'A)^{-1} A' \text{ var } x \, A (A'A)^{-1}$$
$$= \sigma^2 (A'A)^{-1}.$$

Now we can calculate an unbiased estimator of σ^2 by the previous section, namely,

$$\hat{\sigma}^2 = \frac{1}{n-p}(x - A\hat{\beta})'(x - A\hat{\beta}),$$

and it follows that $\hat{\sigma}^2 (A'A)^{-1}$ is an unbiased estimator of the variance matrix of $\hat{\beta}$. If we are interested in estimating a linear parametric function $c'\beta$, say, then $c'\hat{\beta}$ is a minimum variance unbiased linear estimator of this and

$$\text{var}_\beta c'\hat{\beta} = \hat{\sigma}^2 c' (A'A)^{-1} c.$$

Hence an estimate of the standard deviation of our estimator $c'\hat{\beta}$ is

$$\hat{\sigma}\sqrt{\{c'(A'A)^{-1}c\}},$$

a number which can be calculated from the given observations.

In particular, if we wish to estimate a particular component β_i of β, this is estimated by $\hat{\beta}_i$ and

$$\text{var } \hat{\beta}_i = \sigma^2 \times (i, i)\text{th element of } (A'A)^{-1}$$
while estimated S.D. $\hat{\beta}_i = \hat{\sigma}\sqrt{\{(i, i)\text{th element of } (A'A)^{-1}\}}$.

3.9 Normal theory

As a general rule in statistical theory, the more we are prepared to assume about the probabilistic model underlying observations, the stronger the results we can prove regarding estimators. In the preceding sections of this chapter we have made assumptions about the first and second moments of the error vector ε, but no further assumptions about the form of its distribution. Then we were able to demonstrate that least-squares estimators were best in the class of unbiased *linear* estimators. Suppose that we add the assumption that ε has a normal distribution, so that our model now becomes

$$x = A\beta + \varepsilon,$$

where ε is $N(0, \sigma^2 I_n)$; A is known, and β and σ^2 are unknown. Can we now prove something stronger about least-squares estimators? The answer is yes, and we appeal to the Rao–Blackwell theorem to demonstrate this.

With the additional assumption of normality of errors, we have

$$p(x; \beta, \sigma^2) = (2\pi\sigma^2)^{-\frac{1}{2}n} \exp\left[-\frac{1}{2\sigma^2}(x - A\beta)'(x - A\beta)\right]$$

$$= C(\beta, \sigma^2) \exp\left[-\frac{x'x}{2\sigma^2} + \sum_{i=1}^{p} \frac{\beta_i}{\sigma^2} y_i\right],$$

where $y = A'x$.

Now write $t_i(x) = y_i \quad (i = 1, 2, \ldots, p)$,
$t_{p+1}(x) = x'x$,
and $t(x) = \{t_1(x), t_2(x), \ldots, t_{p+1}(x)\}$.

Also, reparametrize in terms of $\theta = (\theta_1, \theta_2, \ldots, \theta_{p+1})$,

where $\theta_i = \dfrac{\beta_i}{\sigma^2} \quad (i = 1, 2, \ldots, p)$,

and $\theta_{p+1} = -\dfrac{1}{2\sigma^2}$.

Then $p(x; \beta, \sigma^2)$ can be expressed in the form

$$C^*(\theta) \exp\left[\sum_{i=1}^{p+1} \theta_i t_i(x)\right].$$

It follows from the factorization theorem that $t(x)$ is sufficient for θ and from Theorem 2.5.4 on exponential families that the family of distributions of t is complete, if there are no prior restrictions on β and σ^2, because then the parameter space contains a $(p+1)$-dimensional rectangle.

Now β_i is a real function of θ, $\beta_i = -\frac{1}{2}\theta_i/\theta_{p+1}$, and $\hat{\beta}_i$ is a function of the sufficient statistic t, since

$$\hat{\beta} = (A'A)^{-1}A'x = (A'A)^{-1}y.$$

Moreover $\hat{\beta}_i$ is an unbiased estimator of β_i. (We assume here that A has full rank so that β_i is estimable.) It follows that $\hat{\beta}_i$ has minimum variance in the class of *all* unbiased estimators of β_i ($i = 1, 2, \ldots, p$), by the argument of section 2.6. Incidentally, if $s^2 = (x - A\hat{\beta})'(x - A\hat{\beta})$, then $s^2/(n-p)$ is a minimum-variance unbiased estimator of σ^2 in this case.

Thus by adding the assumption of normality to the linear model we are able to establish that least-squares estimates are optimal in a stronger sense than they are without this assumption.

3.9.1 Note

It is convenient at this stage to prove, for subsequent use, a result concerning the distributions of $\hat{\beta}$ and s^2 when the assumption of normality of the error ε is added to the linear model assumptions previously adopted.

Since $\hat{\beta}$ is linearly related to a normal random vector ($\hat{\beta} = (A'A)^{-1}A'x$, where x is $N(A\beta, \sigma^2 I)$) we can state immediately that $\hat{\beta}$ itself is

$N\{\beta, \sigma^2 (A'A)^{-1}\}.$

Furthermore we have seen in section 3.4 that there exists an orthogonal matrix P such that if $\eta = P\varepsilon$,

then $\quad PA(\hat{\beta} - \beta) = (\eta_1, \eta_2, \ldots, \eta_p, 0, 0, \ldots, 0)'$
and $\quad P(x - A\hat{\beta}) = (0, 0, \ldots, 0, \eta_{p+1}, \ldots, \eta_n)'$:

Now since the components of ε are independent $N(0, \sigma^2)$ random variables and since P is orthogonal, it follows that $\eta_1, \eta_2, \ldots, \eta_n$ are also independent $N(0, \sigma^2)$. Therefore $\hat{\beta} - \beta$ and $x - A\hat{\beta}$ are independent. Furthermore

$$(x - A\hat{\beta})'(x - A\hat{\beta}) = \sum_{i=p+1}^{n} \eta_i^2$$

and so $s^2 = (x - A\hat{\beta})'(x - A\hat{\beta})$ is distributed as $\sigma^2 \chi^2(n-p)$.

In other words, *with the normal assumption the least-squares estimates $\hat{\beta}_1, \hat{\beta}_2, \ldots, \hat{\beta}_p$ of $\beta_1, \beta_2, \ldots, \beta_p$ respectively are jointly normally distributed and are independent of the residual sum of squares s^2 which is distributed as $\sigma^2 \chi^2 (n-p)$.*

3.10 Least squares with side conditions

Until now in this chapter we have considered the linear model with $E(x)$

expressed in the form $A\beta$. Sometimes the natural expression of the model in terms of the parameters of interest does not occur in this way. In particular these parameters may be mathematically related to one another and often the relationships between them are linear. In the latter case we have a model in which $E(x)$ is expressed in the form $A\beta$ with β_is satisfying certain side conditions, say $H\beta = 0$, where H is a $q \times p$ matrix ($q < p$) of known coefficients. It must be emphasized straightway that from a theoretical point of view this new model is not different in essence from that which we have been discussing. In both cases we are stating that $E(x)$ belongs to a subspace of R^n, and indeed by reparametrization we can throw the new model into the form of the previous one. Suppose, for instance, that rank $H = q$. (If rank $H < q$ this simply means that some of the conditions are redundant, being consequences of the rest, and we may simply discard these.) By adjoining $p-q$ suitable chosen rows to the matrix H we can construct a non-singular $p \times p$ matrix K with $K' = (H', H^{*'})$, say. Now let $\gamma = K\beta$. The side conditions on β are, in terms of γ,

$$\gamma_1 = \gamma_2 = \ldots = \gamma_q = 0.$$

Now we have

$$E(x) = AK^{-1}\gamma = B\gamma, \quad \text{say},$$
$$= (B_1 \, B_2) \begin{bmatrix} \gamma^{(1)} \\ \gamma^{(2)} \end{bmatrix},$$

where $\gamma^{(1)} = (\gamma_1, \gamma_2, \ldots, \gamma_q)'$ and $\gamma^{(2)} = (\gamma_{q+1}, \ldots, \gamma_p)'$.

Thus $E(x) = B_1 \gamma^{(1)} + B_2 \gamma^{(2)} = B_2 \gamma^{(2)}$, when $\gamma^{(1)} = 0$. In this expression for $E(x)$, the side conditions are incorporated, there are no conditions on $\gamma^{(2)}$ and the model is as previously.

In practice, while it would be possible to treat the problem of least-squares estimation with side conditions in the way just described, to determine $\hat{\gamma}$ and then $\hat{\beta} = K^{-1}\hat{\gamma}$, this would be an unnatural approach. The problem we have is that of minimizing the sum of squares

$$(x - A\beta)'(x - A\beta)$$

subject to the side conditions $H\beta = 0$, since it is natural to require that our least-squares estimates should satisfy the conditions which we know to be satisfied by the parameters being estimated. The obvious way of going about this is to introduce Lagrange multipliers and derive the following equations satisfied by the restricted least-squares estimate $\hat{\beta}$. In these equations, λ is a q-vector of Lagrange multipliers, $\lambda_1, \lambda_2, \ldots, \lambda_q$.

$$A'A\hat{\beta} + H'\lambda = A'x,$$
$$H\hat{\beta} = 0,$$

or $\begin{bmatrix} A'A & H' \\ H & 0 \end{bmatrix} \begin{bmatrix} \hat{\beta} \\ \lambda \end{bmatrix} = \begin{bmatrix} A'x \\ 0 \end{bmatrix}.$

The vector λ here obviously depends in general on x and so we may regard it as a random vector.

We now consider the question of what we can say about the distribution of the restricted least-squares estimate $\hat{\beta}$ when we retain the standard assumptions regarding the distribution of the error vector ε, namely that $E(\varepsilon) = 0$ and $\text{var } \varepsilon = \sigma^2 I$, and when $E(x) = A\beta$ where $H\beta = 0$. This question is of interest *per se* and it is also relevant in a problem which will concern us later.

3.10.1 rank A = p, rank H = q

The first case which we shall discuss is that in which there are no identifiability difficulties regarding β (rank $A = p$) and in which no restrictions are redundant (rank $H = q$). In this case $A'A$ is positive definite, and the matrix

$$\begin{bmatrix} A'A & H' \\ H & 0 \end{bmatrix}$$

is non-singular (see Appendix A). Moreover, if its inverse, similarly partitioned, is

$$\begin{bmatrix} P & Q \\ Q' & R \end{bmatrix}$$

then $\begin{bmatrix} \hat{\beta} \\ \lambda \end{bmatrix} = \begin{bmatrix} PA'x \\ Q'A'x \end{bmatrix}$.

Also $E_\beta \begin{bmatrix} \hat{\beta} \\ \lambda \end{bmatrix} = \begin{bmatrix} PA'A\beta \\ Q'A'A\beta \end{bmatrix} = \begin{bmatrix} \beta \\ 0 \end{bmatrix}$,

since $PA'A + QH = I$
and $Q'A'A + RH = 0$,

so that $PA'A\beta = \beta - QH\beta = \beta$,
and $Q'A'A\beta = -RH\beta = 0$,

as $H\beta = 0$.

Furthermore $\text{var}_\beta \begin{bmatrix} \hat{\beta} \\ \lambda \end{bmatrix} = \sigma^2 \begin{bmatrix} PA'AP & PA'AQ \\ Q'A'AP & Q'A'AQ \end{bmatrix}$

$$= \sigma^2 \begin{bmatrix} P & 0 \\ 0 & -R \end{bmatrix},$$

since in addition to the previous matrix equations we have $HP = 0$ and $HQ = I$, and therefore $PA'AP = QHP = P$, etc. (see Appendix A).

To sum up: With the linear model

$x = A\beta + \varepsilon$,

where $H\beta = 0$, $E(\varepsilon) = 0$ and $\text{var } \varepsilon = \sigma^2 I$, rank $A = p$ and rank $H = q$, the restricted least-squares estimate $\hat{\beta}$ has mean β and variance matrix P, the leading $p \times p$ sub-matrix of the inverse of $\begin{bmatrix} A'A & H' \\ H & 0 \end{bmatrix}$.

It is clear from the argument at the beginning of section 3.10 which reparametrizes in terms of γ that

$$\frac{1}{n-p+q}(x-A\hat{\beta})'(x-A\hat{\beta})$$

is in this case an unbiased estimator of σ^2.

3.10.2 rank $A < p$, rank $H = q$

As we have seen, when rank $A < p$ the parameter β is not identifiable and the domain of β is partitioned into equivalence classes of parameters. Any two parameters in the same equivalence class yield the same value for $E(x)$, and we cannot hope to distinguish between these as a result of observation. Indeed in this linear situation it is not difficult to identify these equivalence classes. All parameters β in the null space of A, that is all βs such that $A\beta = 0$ are in the same equivalence class and this is a linear subspace of R^p. Any equivalence class is a 'hyperplane parallel to this subspace'.

One method of proceeding in this case is to introduce restrictions on β in order to focus attention on exactly one member of each equivalence class and to behave as if the true parameter satisfied these restrictions. To take a trivial illustration, suppose that our model specifies that for $i = 1, 2, \ldots, n$, $E(x_i) = \beta_1 + \beta_2$,

i.e., that $E(x) = \begin{bmatrix} 1 & 1 \\ 1 & 1 \\ \vdots & \vdots \\ 1 & 1 \end{bmatrix} \begin{bmatrix} \beta_1 \\ \beta_2 \end{bmatrix}$.

All parameters β such that $\beta_1 + \beta_2$ has a given value k, say, are equivalent. If now we impose the restriction that $\beta_1 = \beta_2$, this restriction serves to pick out exactly one member of each equivalence class – the member $\begin{bmatrix} \frac{1}{2}k \\ \frac{1}{2}k \end{bmatrix}$ of the equivalence class defined by $\beta_1 + \beta_2 = k$. Then we may proceed to estimate β as if it satisfied the restriction $\beta_1 = \beta_2$, and so estimate the equivalence class to which it belongs.

In general when A has order $n \times p$ and rank $r < p$ it is possible to introduce $p-r$ linear restrictions which serve, as in the illustration, to identify a particular member of each equivalence class. More specifically, there exists a $(p-r) \times p$ matrix L of rank $(p-r)$ such that the equations

$A\beta = k$,
$L\beta = 0$

The Method of Least Squares

have exactly one solution. An obvious necessary and sufficient condition for these equations to have a unique solution is that

$$\text{rank} \begin{bmatrix} A \\ L \end{bmatrix} = p,$$

and again obviously we can find many matrices L of order $(p-r) \times p$ which satisfy this condition. Since the conditions $L\beta = 0$ serve to identify β we shall refer to them as identifiability restraints.

It often happens in practice that the natural or symmetric specification of $E(x)$ in a linear model takes the following form:

$$E(x) = A\beta \text{ where } H\beta = 0; \text{ rank } A = r < p \text{ and rank } H = q.$$

Moreover some of the side conditions $H\beta = 0$ serve to identify β and the remainder are 'genuine' restrictions on β. In fact there exists a sub-matrix of H, H_1 say, of order $(p-r) \times p$ such that $\begin{bmatrix} A \\ H_1 \end{bmatrix}$ has rank p. Again in theory this specification of the linear model presents no essentially new difficulty for by a reparametrization we can clearly revert to the original form of the model which we have discussed in detail. However we now consider the practical algebra corresponding to this specification.

We may suppose without any loss of generality that H may be partitioned into $\begin{bmatrix} H_1 \\ H_2 \end{bmatrix}$,

where H_1 has $p-r$ rows and the equations $H_1\beta = 0$ are identifiability constraints, so that $\begin{bmatrix} A \\ H_1 \end{bmatrix}$ has rank p, and $A'A + H_1'H_1$ is a $p \times p$ positive definite matrix. As before, the restricted least-squares estimate $\hat{\beta}$ satisfies the equations

$$\begin{bmatrix} A'A & H' \\ H & 0 \end{bmatrix} \begin{bmatrix} \hat{\beta} \\ \lambda \end{bmatrix} = \begin{bmatrix} A'x \\ 0 \end{bmatrix}.$$

But now $A'A$ is singular and some modification of the argument of section 3.8.1 is necessary. This modification is relatively simple. For since $H\hat{\beta} = 0$, so that in particular $H_1\hat{\beta} = 0$, an equivalent set of equations is

$$\begin{bmatrix} A'A + H_1'H_1 & H' \\ H & 0 \end{bmatrix} \begin{bmatrix} \hat{\beta} \\ \lambda \end{bmatrix} = \begin{bmatrix} A'x \\ 0 \end{bmatrix},$$

and the matrix $A'A + H_1'H_1$ is positive definite, so that we now have a set of equations similar in structure to those of section 3.8.1; and the matrix on the left hand side is non-singular. If now we set

$$\begin{bmatrix} A'A + H_1'H_1 & H' \\ H & 0 \end{bmatrix}^{-1} = \begin{bmatrix} P & Q \\ Q' & R \end{bmatrix}$$

then we have, as before

$$\hat{\beta} = PA'x$$
and $\quad \lambda = Q'A'x.$

The matrix relationships used to establish the results of section 3.8.1 were

$PA'A + QH = I$	3.1
$Q'A'A + RH = 0$	3.2
$HP = 0$	3.3
$HQ = I$	3.4

and these are now replaced by the following

$P(A'A + H'_1 H_1) + QH = I$	3.1a
$Q'(A'A + H'_1 H_1) + RH = 0$	3.2a
$HP = 0$	3.3a
$HQ = I$	3.4a

From equation **3.3a** we have in particular the fact that $PH'_1 = 0$, so equation **3.1a** is equivalent to equation **3.1** and the only essential difference between the second and first set of equations is the term $Q'H'_1 H_1$ in equation **3.2a**. As is easily verified the only difference that this makes to the deductions of section 3.8.1 is that now

$$\text{var } \lambda = -R - Q'H'_1 H_1 Q;$$

everything else remains unaltered. Since $HQ = I$ and $p - r < q$ it follows that

$$(H_1 Q)'(H_1 Q) = \begin{bmatrix} I_{p-r} & 0 \\ 0 & 0 \end{bmatrix},$$

so the only adjustments required by the non-identifiability of β are that
(a) we replace $A'A$ by $A'A + H'_1 H_1$,
(b) var λ becomes $-R - \begin{bmatrix} I_{p-r} & 0 \\ 0 & 0 \end{bmatrix}$ instead of, as previously $-R$.

While we are not particularly interested in the random variable λ in the meantime, these results concerning λ which emerge here in a natural way, will be used in a subsequent problem.

3.11 **Discussion**

There are many variations on the least-squares theme and there are various questions which we have left unanswered. The method is an extremely useful one and it is often applied even when the assumptions of the Gauss–Markov theorem, which justifies it in terms of minimum-variance unbiasedness, are not satisfied. It therefore becomes natural to inquire how the properties of the method are affected by changes in the assumptions regarding the error vector etc. Much of econometrics is concerned with this kind of question and for a very full discussion of such points the reader is referred to Malinvaud (1966).

Examples

3.1 Assume that observations $x_1, x_2, \ldots, x_n$ can be expressed in the form
$$x_i = \beta_0 + \beta_1 a_i + \varepsilon_i \quad (i = 1, 2, \ldots, n),$$
where $a_1, a_2, \ldots, a_n$ are known values of a concomitant variable and the εs are uncorrelated errors with common variance σ^2. Verify that β_0 and β_1 are both estimable if and only if the a_is are not all equal and confirm the intuitive acceptability of this result by imagining a scatter diagram of the points (a_i, x_i). Show that, when the a_is are not all equal, least-squares estimates $\hat{\beta}_0$ and $\hat{\beta}_1$ are given by
$$\hat{\beta}_1 = \frac{\sum (a_i - \bar{a}) x_i}{\sum (a_i - \bar{a})^2},$$
$$\hat{\beta}_0 = \bar{x} - \bar{a} \hat{\beta}_1.$$
Prove directly from the first expression that $\operatorname{var} \hat{\beta}_1 = \sigma^2 / \sum (a_i - \bar{a})^2$. Show that $\bar{x}$ and $\hat{\beta}_1$ have zero covariance and deduce that

(a) $\operatorname{cov}(\hat{\beta}_0, \hat{\beta}_1) = -\bar{a} \operatorname{var} \hat{\beta}_1$;

(b) $\operatorname{var} \hat{\beta}_0 = \sigma^2 \left[\dfrac{1}{n} + \dfrac{\bar{a}^2}{\sum (a_i - \bar{a})^2} \right].$

Verify these results by writing the model in the matrix notation
$$x = A\beta + \varepsilon$$
and using the general results of chapter 3.

3.2 Observations $x_1, x_2, \ldots, x_n$ can be expressed in the form
$$x_i = \beta_0 + \beta_1 a_i + \beta_2 a_i^2 + \varepsilon_i \quad (i = 1, 2, \ldots, n),$$
where the a_is are values of a concomitant variable and the εs are uncorrelated errors with common variance. Establish that $\beta = (\beta_0, \beta_1, \beta_2)$ is identifiable if and only if there are at least three different values among $a_1, a_2, \ldots, a_n$.

3.3 The model $x_i = \beta_0 + \beta_1 a_i + \varepsilon_i \quad (i = 1, 2, \ldots, n),$
may be expressed in the form
$$x_i = \beta_0 + \beta_1 \bar{a} + \beta_1 (a_i - \bar{a}) + \varepsilon_i$$
$$= \alpha + \beta_1 (a_i - \bar{a}) + \varepsilon_i, \quad \text{say}.$$
Show that this reparametrization in terms of α and β_1 rather than β_0 and β_1 facilitates calculation of least-squares estimates.

Verify that, in general, the model

$$x = A\beta + \varepsilon$$

can always, by reparametrization, be expressed in the form

$$x = B\gamma + \varepsilon,$$

where B is a matrix whose columns are orthogonal, and that the least-squares estimate of γ is easily calculated.

3.4 Observations x_{ij} ($i = 1, 2, \ldots, r$; $j = 1, 2, \ldots, n$), are such that

$$x_{ij} = \mu + \tau_i + \varepsilon_{ij},$$

where the εs are uncorrelated errors with a common variance. Verify that $\mu, \tau_1, \tau_2, \ldots, \tau_r$ are not identifiable, but that they are when the restriction $\tau_1 + \tau_2 + \ldots + \tau_r = 0$ is imposed. Show that the least-squares estimates, subject to this restriction are

$$\hat{\mu} = x_{..} = \frac{1}{rn} \sum_{i,j} x_{ij},$$

$$\hat{\tau}_i = x_{i.} - x_{..},$$

where $x_{i.} = \dfrac{1}{n} \sum_j x_{ij}$ ($i = 1, 2, \ldots, r$).

3.5 Aerial observations x_1, x_2, x_3, x_4 are made of the angles $\theta_1, \theta_2, \theta_3, \theta_4$ of a quadrilateral on the ground. If these observations are subject to independent errors with zero means and common variance σ^2, determine least-squares estimates of the θs, and obtain an unbiased estimate of σ^2.

Suppose that the quadrilateral is known to be a parallelogram with $\theta_1 = \theta_3$ and $\theta_2 = \theta_4$. What then are the least-squares estimate of its angles, and how would you estimate σ^2?

3.6 A chemical compound can be produced by a certain process without the help of a catalyst, but it is hoped that the yield will be increased if a catalyst is present. To investigate this, five identical containers are used in the following way.

Container	Treatment	Yield
1	No catalyst	x_1
2	Catalyst A at strength a_1	x_2
3	Catalyst A at strength a_1	x_3
4	Catalyst B at strength a_2	x_4
5	Catalyst B at strength $2a_2$	x_5

Assuming that regression of yield on strength is linear for each catalyst, obtain least-squares estimates of the 'unaided' effect and of the two regression coefficients.

Derive the variance matrix of these estimators (making the usual assumptions about errors) and deduce that, for given a_1, the least-squares estimator of the difference of regression coefficients has minimum variance when $a_2 = a_1$.

3.7 A deterministic process $y_0, y_1, \ldots, y_n$ is governed by the relation

$$y_{i+1} = a y_i \quad (i = 0, 1, \ldots, n-1),$$

where a is a known constant. The y_is cannot be observed without error and observations $x_0, x_1, \ldots, x_n$ are such that

$$x_i = y_i + \varepsilon_i \quad (i = 0, 1, \ldots, n),$$

where the ε_is are uncorrelated errors with common variance. Determine least-squares estimates of $y_0, y_1, \ldots, y_n$.

If a were unknown, how then would you estimate $y_0, y_1, \ldots, y_n$?

In each case obtain an unbiased estimate of the error variance.

3.8 In example 3.4, take $r = 3$ and verify directly the general results of section 3.10.2.

4 The Method of Maximum Likelihood

4.1 The likelihood function

The justification of the method of least squares requires no knowledge of the form of the distribution of the error vector apart from its mean and variance matrix, and the method can be applied without this further knowledge. The method of maximum likelihood, on the other hand, is applicable mainly in situations where the true distribution on the sample space is known apart from the values of a finite number of unknown real parameters. So maximum likelihood is usually applied when the family of possible distributions on the sample space can be labelled by a parameter θ taking values in a finite-dimensional Euclidean space. In addition, its application is generally restricted to the case where this family $\{P_\theta : \theta \in \Theta\}$ (Θ a subset of R^s, say) possesses density functions $\{p_\theta ; \theta \in \Theta\}$ with respect to some 'natural' measure on the sample space, such as counting measure if the sample space is discrete or Lebesgue measure when it is not; in the discrete case $p_\theta(x)$ is 'the probability of the point x when θ is the true parameter'; in the continuous case $p_\theta(x)$ is 'the probability density at x when θ is the true parameter'.

It is convenient now to change our notation and write $p(x, \theta)$ instead of $p_\theta(x)$; and we make a distinction between the function $p(\cdot, \theta)$ which is a density function on the sample space, and the function $p(x, \cdot)$ which is a function on the parameter space. The latter function, $p(x, \cdot)$, is called the *likelihood function* corresponding to the observation x, or simply the likelihood function. It expresses the plausibilities of different parameters after we have observed x, in the absence of any other information we may have about these different values. (This last sentence might well be the subject of some controversy, but we shall return to this point later.)

The method of maximum likelihood has a strong intuitive appeal and according to it, we estimate the true parameter θ by any parameter which maximizes the likelihood function $p(x, \cdot)$; such a parameter belongs to the set most plausible after we have observed x. Often there is a unique maximizing parameter which is *the* most plausible and this is then *the* maximum-likelihood estimate.

4.1.1 *Definition*

A maximum-likelihood estimate $\hat{\theta}(x)$ is any element of Θ such that

$$p\{x, \hat{\theta}(x)\} = \max_{\theta \in \Theta} p(x, \theta).$$

Of course it is possible, if, for instance, Θ is an open set, that no maximum-likelihood estimate exists. However in practice this does not often cause trouble.

Again formally at this stage we make the distinction between the estimate $\hat{\theta}(x)$ and the estimator $\hat{\theta}$, but we shall not maintain this distinction consistently, leaving the context to make it clear whether we are thinking of $\hat{\theta}(x)$ as a function or as a particular value of a function.

4.1.2 Example

The results of n independent trials in each of which the probability of success is θ are $x = (x_1, x_2, \ldots, x_n)$, where as usual each x_i is either 0 or 1. Find the maximum-likelihood estimate of θ.

The likelihood function, defined on the interval $(0, 1)$, is given by

$$p(x, \theta) = \theta^{\Sigma x_i}(1-\theta)^{n-\Sigma x_i},$$

and its maximum occurs at

$$\hat{\theta}(x) = \frac{\sum x_i}{n} = \frac{\text{number of successes}}{\text{number of trials}}.$$

So in this case the maximum-likelihood estimator coincides with the M.V.U.E.

4.1.3 Example

Let $x = (x_1, x_2, \ldots, x_n)$ be a random sample from an $N(\mu, \sigma^2)$ distribution with μ and σ^2 unknown. Find maximum likelihood estimates of μ and σ^2 or, equivalently, the maximum likelihood estimate of $\theta = (\mu, \sigma^2)$. In this case the likelihood function, defined for all real μ and all $\sigma^2 > 0$, is

$$p(x, \mu, \sigma^2) = \frac{1}{\sigma^n (2\pi)^{\frac{1}{2}n}} \exp\left[-\frac{1}{2\sigma^2} \sum (x_i - \mu)^2\right].$$

Maximizing p, which is non-negative, is equivalent to maximizing $\log p$ and

$$\log p(x, \mu, \sigma^2) = \text{constant} - n \log \sigma - \frac{1}{2\sigma^2} \sum (x_i - \mu)^2.$$

We find the maximizing values by the standard method for maximizing a function of two variables, namely equating partial derivatives to zero. This gives

$$\frac{1}{\sigma^2} \sum (x_i - \mu) = 0$$

and

$$-\frac{n}{\sigma} + \frac{1}{\sigma^3} \sum (x_i - \mu)^2 = 0,$$

equations which have the unique solution

$$\hat{\mu} = \bar{x} = \frac{1}{n}\sum x_i \quad \text{and} \quad \hat{\sigma}^2 = \frac{1}{n}\sum (x_i - \bar{x})^2.$$

It is not difficult to verify that these values of μ and σ^2 yield an absolute (not only a local) maximum of the log-likelihood, so that they are maximum-likelihood estimates.

4.2 Calculation of maximum-likelihood estimates

In these two examples it was possible to find relatively simple expressions in closed form for maximum-likelihood estimates, but often this is not possible and numerical methods are necessary. It is usually possible to assume that maximum-likelihood estimates emerge as a solution of the 'likelihood equations', namely

$$\frac{\partial}{\partial \theta_i} \log p(x, \theta) = 0, \quad (i = 1, 2, \ldots, s).$$

However, these equations often have to be solved numerically.

A standard method of solving the likelihood equations is Newton's method or an adaptation of it. Symbolically the equations we have to solve may be written

$$D_\theta l(x, \theta) = 0,$$

where $l(x, \theta) = \log p(x, \theta)$ and D_θ is the vector differential operator whose ith component is $\partial/\partial \theta_i$. By exploiting special features of the situation under investigation, as in the example following this section, we can often obtain a good initial approximation $\theta^{(0)}$ to the solution $\hat{\theta}$ of these equations. Then we expand by Taylor's theorem as far as terms of first order in $\hat{\theta} - \theta^{(0)}$ to obtain

$$0 = D_\theta l(x, \hat{\theta}) \simeq D_\theta l(x, \theta^{(0)}) + \{D_\theta^2 l(x, \theta^{(0)})\}(\hat{\theta} - \theta^{(0)}),$$

where D_θ^2 is the matrix operator

$$\left[\frac{\partial^2}{\partial \theta_i \, \partial \theta_j}\right].$$

It follows from this that

$$\hat{\theta} \simeq \theta^{(0)} - \{D_\theta^2 l(x, \theta^{(0)})\}^{-1} D_\theta l(x, \theta^{(0)}),$$

and the right hand side of this equation is a new approximation $\theta^{(1)}$ to the maximum-likelihood estimate $\hat{\theta}$.

Now we repeat this process, using $\theta^{(1)}$ instead of $\theta^{(0)}$, to obtain a new approximation $\theta^{(2)}$, and so on. Thus we establish an iterative procedure for obtaining a sequence $(\theta^{(n)})$ which usually converges to $\hat{\theta}$. This is Newton's method.

The laborious aspect of this iterative procedure is the inversion of the matrix $D_\theta^2 l(x, \theta^{(i)})$ at the ith stage. If our initial approximation $\theta^{(0)}$ is good, then $D_\theta^2 l(x, \theta^{(0)})$ will be near $D_\theta^2 l(x, \theta^{(i)})$ in non-pathological conditions, so that we can often use the former matrix *at each stage* of the procedure and so avoid the necessity for a new matrix inversion at every stage. This modified procedure leads to a new sequence of approximations to $\hat{\theta}$, a sequence which usually converges to $\hat{\theta}$, though possibly more slowly than the sequence $(\theta^{(n)})$.

A further modification sometimes reduces the total amount of computation even further. There is sometimes good reason to suppose that the matrix $D_\theta^2 l(x, \theta^{(0)})$ will be relatively close to its expected value $E_\theta^{(0)} \{D_\theta^2 l(x, \theta^{(0)})\}$; close enough, that is, to ensure that a sequence of approximations to $\hat{\theta}$, based on the use of this expected value rather than on $D_\theta^2 l(x, \theta^{(0)})$ itself, will still converge to $\hat{\theta}$. Now it often happens that terms awkward to calculate appear in $D_\theta^2 l(x, \theta^{(0)})$ but not in its expected value. So again it is sometimes possible to reduce total calculation by using $E_\theta^{(0)} \{D_\theta^2 l(x, \theta^{(0)})\}$ in place of $D_\theta^2 l(x, \theta^{(0)})$.

We recall that $E_\theta^{(0)} \{D_\theta^2 l(x, \theta^{(0)})\}$ is simply, in most instances, $-B_\theta^{(0)}$ where B_θ is the information matrix (see section 2.13). It follows that the fully modified iterative procedure is defined as follows:

$$\theta^{(n+1)} = \theta^{(n)} + B_{\theta^{(0)}}^{-1} \{D_\theta l(x, \theta^{(n)})\},$$

where $\theta^{(0)}$ is an initial approximation to $\hat{\theta}$, usually obtained by exploiting special statistical features of the problem involved.

4.2.1 Example

Suppose that it may be assumed that the probability $\pi(s)$ that an individual responds to the level s of a stimulus can be expressed, at least approximately, in the form

$$\pi(s) = \Phi\left(\frac{s-\mu}{\sigma}\right) = \frac{1}{\sqrt{(2\pi)}} \int_{-\infty}^{(s-\mu)/\sigma} e^{-\frac{1}{2}z^2} \, dz,$$

an assumption which may appear somewhat drastic, but which in fact turns out to be valid in many circumstances. The level s_i of the stimulus is applied to n_i individuals ($i = 1, 2, \ldots, r$) and the numbers m_i ($i = 1, 2, \ldots, r$) of responses at the different levels are observed. Determine maximum-likelihood estimates of μ and σ.

In this particular case we have, in our general notation,

$$x = (m_1, m_2, \ldots, m_r) \qquad \theta = (\mu, \sigma)$$

and $\quad p(x, \theta) = \prod_{i=1}^{r} \binom{n_i}{m_i} \{\pi(s_i)\}^{m_i} \{1 - \pi(s_i)\}^{n_i - m_i},$

assuming, of course, that individuals respond independently of one another.

Hence, writing π_i in place of $\pi(s_i)$ for symmetry of notation, we have

$$l(x, \theta) = \text{constant} + \sum_i \{m_i \log \pi_i + (n_i - m_i) \log (1 - \pi_i)\},$$

and the likelihood equations, $D_\theta l(x, \theta) = 0$, are

$$\frac{\partial}{\partial \mu} l(x, \theta) = \sum_i \frac{m_i - n_i \pi_i}{\pi_i (1 - \pi_i)} \frac{\partial \pi_i}{\partial \mu} = 0$$

and $$\frac{\partial}{\partial \sigma} l(x, \theta) = \sum_i \frac{m_i - n_i \pi_i}{\pi_i (1 - \pi_i)} \frac{\partial \pi_i}{\partial \sigma} = 0.$$

It will be appreciated that these equations are not susceptible to methods of solution which are other than numerical, and our first problem is to obtain initial approximations μ_0 and σ_0 to their solution.

Φ is a monotonic increasing function. Let Φ^{-1} denote the inverse function defined on $(0, 1)$. If we knew the π_is, then when we plotted the points $(s_i, \Phi^{-1}(\pi_i))$, according to our assumption regarding π, these points would lie on the straight line

$$\Phi^{-1}(\pi) = \frac{(s - \mu)}{\sigma}.$$

Of course we do not know the π_is but we do have estimates of them since m_i/n_i is an estimate of $\pi_i (i = 1, 2, \ldots, r)$. Consequently if we plot the points $(s_i, \Phi^{-1}(m_i/n_i))$, and if our assumption regarding π is justified, these points should be scattered around a straight line. Plotting these points therefore gives us at the same time a check on the validity of our assumption about π (if they are obviously non-linear the assumption is not justified), and a means of obtaining initial approximations to the solution of the likelihood equations. For if we fit a straight line to this set of points, the parameters of the fitted line yield estimates of the true parameters μ and σ, estimates which approximate to the maximum-likelihood estimates, the solution of the likelihood equations.

We now illustrate the point of replacing $D_\theta^2 l(x, \theta)$ by its expected value. In our example a typical element of the former matrix is

$$\frac{\partial^2}{\partial \mu^2} l(x, \theta),$$

which is rather a complicated expression. Note however that all but one of the terms which arise under the summation sign when we differentiate $\partial l(x, \theta)/\partial \mu$ with respect to μ, contain as a factor $m_i - n_i \pi_i$, whose expectation is zero. It follows that

$$E_\theta \left[\frac{\partial^2}{\partial \mu^2} l(x, \theta) \right] = \sum_i \frac{-n_i}{\pi_i (1 - \pi_i)} \left(\frac{\partial \pi_i}{\partial \mu} \right)^2,$$

and similarly that the information matrix B_θ is given by

$$B_\theta = \begin{bmatrix} \sum_i \dfrac{n_i}{\pi_i(1-\pi_i)}\left(\dfrac{\partial \pi_i}{\partial \mu}\right)^2 & \sum_i \dfrac{n_i}{\pi_i(1-\pi_i)}\dfrac{\partial \pi_i}{\partial \mu}\dfrac{\partial \pi_i}{\partial \sigma} \\ \sum_i \dfrac{n_i}{\pi_i(1-\pi_i)}\dfrac{\partial \pi_i}{\partial \mu}\dfrac{\partial \pi_i}{\partial \sigma} & \sum_i \dfrac{n_i}{\pi_i(1-\pi_i)}\left(\dfrac{\partial \pi_i}{\partial \sigma}\right)^2 \end{bmatrix}.$$

With our assumption regarding the form of π_i we have

$$\frac{\partial \pi_i}{\partial \mu} = -\frac{1}{\sigma}\phi\left(\frac{s_i - \mu}{\sigma}\right), \quad \frac{\partial \pi_i}{\partial \sigma} = -\frac{s_i - \mu}{\sigma^2}\phi\left(\frac{s_i - \mu}{\sigma}\right),$$

where $\phi(y) = \dfrac{d}{dy}\Phi(y)$.

The calculations involved in the iterative procedure for evaluating $\hat{\mu}$ and $\hat{\sigma}$ are still not trivial, but they are not prohibitive. For further details of the organization of these calculations, and for numerical examples, the reader may refer to Finney (1947); this is an example of an important practical technique called *probit analysis*.

4.3 Optimal properties of maximum-likelihood estimators

The Gauss–Markov theorem provides a justification for the method of least-squares in terms of the concept of minimum-variance unbiasedness, and it is natural to inquire whether a similar justification for the method of maximum likelihood can be found. Unfortunately it is not generally true that maximum-likelihood estimators are unbiased; for instance if $x_1, x_2, \ldots, x_n$ is a random sample from an $N(\mu, \sigma^2)$ distribution with μ and σ^2 unknown, the maximum-likelihood estimator of $\theta = (\mu, \sigma^2)$ is $(\bar{x}, s^2)$

where $s^2 = \dfrac{1}{n}\sum (x_i - \bar{x})^2$,

and while it is true that $E_\theta(\bar{x}) = \mu$, for all θ, it is *not* true that $E_\theta(s^2) = \sigma^2$. In fact $E_\theta(s^2) = (n-1)n^{-1}\sigma^2$, for all θ. (Of course whether we use this as a criticism of the method of maximum likelihood or as a criticism of the concept of unbiasedness is a moot point.)

We can make one or two fairly obvious statements which provide a very partial justification of the method.

Firstly in a 'regular' situation where there exists an unbiased estimator whose variance attains the Cramér–Rao lower bound, the maximum-likelihood estimator coincides with this. For then (section 2.10.1) $\partial \log p(x, \theta)/\partial \theta$ can be expressed in the form $a(\theta)\{\tilde{\theta}(x) - \theta\}$, and the only solution of the likelihood equation $\partial \log p(x, \theta)/\partial \theta = 0$ is $\theta = \tilde{\theta}(x)$, which gives an absolute maximum of $\log p(x, \theta)$ and therefore $\tilde{\theta}(x)$ is the maximum-likelihood estimate.

Secondly it is often possible to show that a maximum-likelihood estimator has high efficiency (section 2.11) in the Fisherian sense. This of course provides a justification only in particular cases.

Thirdly we can say that the maximum-likelihood estimator is a function of a minimal-sufficient statistic. This follows directly from the factorization theorem (section 2.3.3) and it means that the maximum-likelihood estimator depends only on relevant information contained in an observation. It does not mean necessarily that it makes the 'best use' of this information according to some specified definition of 'best use'. The main justification of the method of maximum likelihood is a 'large-sample' one, which shows that when an observation provides lots of information about an unknown parameter, the method utilizes essentially all of this information. We expand this rather vague statement in the following sections.

4.4 Large-sample properties

When we talk about a large sample we mean that the observation x takes the form $x = (x_1, x_2, \ldots, x_n)$, where n is large, and the x_is are independent and identically distributed.

In this case $p(x, \theta) = \prod_{i=1}^{n} p^*(x_i, \theta),$

where $p^*(\cdot, \theta)$ is the density function, corresponding to the parameter value θ, on the space of a 'single observation'.

Also $l(x, \theta) = \log p(x, \theta) = \sum_{i=1}^{n} \log p^*(x_i, \theta),$

regarded as a random variable, is the sum of the independent identically distributed random variables $\log p^*(x_i, \theta)$ $(i = 1, 2, \ldots, n)$.

4.4.1
Now let us fix attention on one particular distribution on the sample space, say that corresponding to the parameter θ_0, which we will think of as the true parameter. For *any* fixed θ, $l(x, \theta)$ is a random variable whose distribution is determined by the 'true' distribution on the sample space.

Let $z(\theta) = E_0 \left[\frac{1}{n} l(x, \theta) \right] = E_0 \{ \log p^*(x_i, \theta) \},$

where the subscript on the expectation operator is used to emphasize the fact that we are taking expectations relative to the distribution corresponding to θ_0.

This function $z(\theta)$ has a property which is, in a sense, the key to the study of large-sample properties of maximum-likelihood estimators: $z(\theta)$ attains its maximum value at θ_0, and if distributions on the sample space corresponding to different parameters are essentially different, then *for no other θ is $z(\theta)$ equal to $z(\theta_0)$*. This important result is a particular case of the following general result derived from Jensen's inequality.

4.4.2 Theorem

Let q and r be density functions of two different probability distributions on the same probability space Y of points y, these distributions being different in the sense that there exists a set of positive q-probability on which $q(y) \neq r(y)$; and let C be any continuous convex function of a non-negative variable.

Then $E_q\left[C\left(\dfrac{r}{q}\right)\right] \geq C(1),$

with strict inequality if C is strictly convex.

Proof. By Jensen's inequality we have

$$E_q\left[C\left(\frac{r}{q}\right)\right] \geq C\left[E_q\left(\frac{r}{q}\right)\right]$$

and the inequality is strict if C is strictly convex, since r/q is not constant with q-probability 1 by assumption.

This simple proof is now completed by the remark that

$$E_q\left(\frac{r}{q}\right) = \int_Y \frac{r(y)}{q(y)} q(y)\, dy, \quad \text{symbolically}$$

$$= \int_Y r(y)\, dy = 1,$$

and so $E_q\left[C\left(\dfrac{r}{q}\right)\right] \geq C(1).$

4.4.3

In the above theorem, let $C = -\log$, let Y be the space of a 'single observation' and let $q = p^*(\cdot, \theta_0)$, $r = p^*(\cdot, \theta)$.

This gives $E_{\theta_0}\left[-\log \dfrac{p^*(\cdot, \theta)}{p^*(\cdot, \theta_0)}\right] \geq -\log 1 = 0$

i.e. $z(\theta_0) - z(\theta) \geq 0$, or $z(\theta_0) \geq z(\theta)$,

and the inequality is strict if the distributions corresponding to θ_0 and θ are essentially different.

So far we have assumed no structure on the parameter space Θ. Typically this space will have a mathematical structure; in particular, if it is a Euclidean space, it has a metric, and then it is usually the case that when θ is near θ_0, $z(\theta_0) - z(\theta)$ is small and when θ is far away from θ_0, $z(\theta_0)$ is considerably larger than $z(\theta)$.

Also in the large sample case the *law of large numbers* ensures that when n is large, $n^{-1} l(x, \theta)$ is, for most x and each particular θ, near $z(\theta)$. Suppose that we assume sufficient regularity to enable us to demonstrate that, for large n and most x, $n^{-1} l(x, \theta)$ is *uniformly* (with respect to θ) near $z(\theta)$. Then the following picture emerges for the case where θ is a real parameter, a case of sufficient generality to illustrate the general case also.

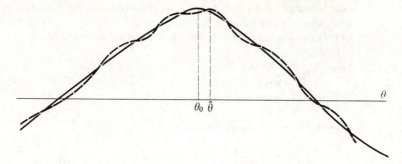

Figure 3 Proximity of the graphs of $z(\theta)$ (unbroken line) and $\frac{1}{n}l(x, \theta)$ (broken line) ensures that $\hat{\theta}$ is near θ_0

In Figure 3, the unbroken curve is the graph of z and the dotted curve is the graph of $n^{-1} l(x, \theta)$ for a typical x. The fact that z assumes its maximum value at θ_0, and that $n^{-1} l(x, \theta)$ is uniformly near $z(\theta)$ ensures that $n^{-1} l(x, \theta)$ assumes its maximum value at a point near θ_0, that is, $\hat{\theta}(x)$ is near θ_0.

4.5 Consistency

In section 4.4 we have outlined the ideas underlying a proof of the fact that the method of maximum likelihood has a property called consistency, defined as follows.

4.5.1 *Definition*

Let $(\tilde{\theta}_n)$ be a sequence of estimators of a parameter θ belonging to a metric space Θ. This sequence is said to be weakly consistent if $\tilde{\theta}_n$ tends in θ-probability to θ; strongly consistent if $\tilde{\theta}_n$ converges with θ-probability one to θ, for all $\theta \in \Theta$.

4.5.2 The reader who is unfamiliar with general metric spaces may think of the parameter space Θ as being the real line, without losing anything essential from the statistical idea here. This idea arises from the following consideration:

Suppose that we continue repeating an experiment which, we feel, is in some sense providing information about an unknown real parameter θ involved in a probabilistic model of the experiment. If the repetitions are independent, then, as their number increases, we feel that we ought to be obtaining more and more information about θ; that, if we are estimating θ, our estimates should get closer and closer to the true value, whatever this may be; and that finally, when the number of repetitions is very, very large, we ought to be fairly certain about what the true value of the parameter is.

Precise mathematical content is given to this notion by the statement that *consistent* estimation should be possible in the circumstances described. It then becomes important to demonstrate that any method of estimation which we employ does have this property of consistency.

Now if (x_n) is a sequence of random variables whose joint distributions depend on an unknown parameter θ in a metric space Θ, we may define a sequence $(\hat{\theta}_n)$ of maximum-likelihood estimators of θ, $\hat{\theta}_n$ being the maximum-likelihood estimator based on $x_1, x_2, \ldots, x_n$. Section 4.4 outlines the main ideas underlying a proof of the fact that if the x_is are independent and identically distributed, the sequence $(\hat{\theta}_n)$ is consistent: weakly consistent if a weak law of large numbers is employed; strongly consistent if a strong law is used. Of course analytic details are required and regularity conditions must be introduced, for a complete proof, which is quite complicated. The reader is referred to Wald (1949) for such a complete proof.

4.6 Large-sample efficiency

The main justification of the method of maximum likelihood in terms of the criterion of minimum-variance unbiasedness is that it is possible to show that for large samples, subject to regularity conditions, maximum-likelihood estimators are nearly unbiased and have variances nearly equal to the Cramér–Rao lower bound. Again a full proof of this result is hedged around with analytic details and regularity conditions and we content ourselves with a heuristic argument. We consider the case where there is an unknown *real* parameter θ, a case of sufficient generality to illustrate the probabilistic content of the argument.

Suppose then that (x_n) is a sequence of independent identically distributed random variables, the distribution of each being known apart from the value of a single real parameter θ; and let $\hat{\theta}_n$ be the maximum-likelihood estimator (which we assume unique) of θ derived from $x_1, x_2, \ldots, x_n$. We shall now assume that n is large and we shall omit the subscript n for typographical brevity. We assume also that $\hat{\theta}$ emerges as a solution of the likelihood equation

$$D_\theta l(x, \theta) = 0,$$

where $x = (x_1, x_2, \ldots, x_n)$,

$$l(x, \theta) = \log p(x, \theta) = \sum_{i=1}^{n} \log p^*(x_i, \theta)$$

and D_θ is the differential operator $\partial/\partial\theta$, as before.

Section 4.4 tells us that with θ-probability near 1, $\hat{\theta}$ is near θ. We therefore have

$$0 = D_\theta l(x, \hat{\theta}) = D_\theta l(x, \theta) + (\hat{\theta} - \theta) D_\theta^2 l(x, \theta) + R(x, \theta, \hat{\theta}),$$

where $R(x, \theta, \hat{\theta})$ is a remainder term involving $(\hat{\theta}-\theta)^2$, which may be shown to be of smaller order than the first-order term, $(\hat{\theta}-\theta)D_\theta^2 l(x, \theta)$, if regularity conditions are satisfied. We can therefore say that, with θ-probability near 1,

$$\hat{\theta}-\theta \simeq -\frac{D_\theta l(x, \theta)}{D_\theta^2 l(x, \theta)}$$

or $\sqrt{n}(\hat{\theta}-\theta) \simeq \frac{n^{-\frac{1}{2}}D_\theta l(x, \theta)}{n^{-1}D_\theta^2 l(x, \theta)}.$

Now $\frac{1}{\sqrt{n}} D_\theta l(x, \theta) = \frac{1}{\sqrt{n}} \sum_{i=1}^n D_\theta \log p^*(x_i, \theta),$

and each random variable in the sum on the right hand side has zero mean and variance i_θ, Fisher's measure of information from a single observation (section 2.9). Consequently by the central limit theorem, $n^{-\frac{1}{2}}D_\theta l(x, \theta)$ is approximately $N(0, i_\theta)$.

Moreover $\frac{1}{n} D_\theta^2 l(x, \theta) = \frac{1}{n} \sum_{i=1}^n D_\theta^2 \log p^*(x_i, \theta)$

and, by Lemma 2.11.2,

$E_\theta\{-D_\theta^2 \log p^*(x_i, \theta)\} = i_\theta.$

Therefore, by a law of large numbers, $-n^{-1}D_\theta^2 l(x, \theta)$ is approximately equal to i_θ.

It follows that $\sqrt{n}(\hat{\theta}-\theta)$ is approximately $i_\theta^{-1} \times$ (an $N(0, i_\theta)$ random variable), so that $\hat{\theta}$ is approximately $N\{\theta, (ni_\theta)^{-1}\}$, i.e., $N(\theta, I_\theta^{-1})$ where I_θ^{-1} is the inverse of Fisher's measure of information from $x_1, x_2, \ldots, x_n$, or the Cramér-Rao lower bound for the variance of an unbiased estimator of θ based on $x_1, x_2, \ldots, x_n$.

So we have 'proved' that maximum-likelihood estimators are efficient for large samples and in addition that $\hat{\theta}$ is approximately normally distributed, in this case where there is a single unknown real parameter. For a complete proof of this result the reader is referred to Cramér (1946), p. 500.

4.6.1 This property generalizes to the case where θ is a vector-valued parameter. The basic results used in the above proof are:

(a) Taylor's theorem in the expansion of $D_\theta l(x, \hat{\theta})$;

(b) a central limit theorem applied to $n^{-\frac{1}{2}}D\, l(x, \theta)$;

(c) a law of large numbers applied to $n^{-1}D^2\, l(x, \theta)$.

Each of these results has a multivariate version and the vector-parameter argument is simply a straight generalization of that above, yielding the result that in this case, for large samples, $\hat{\theta}$ is approximately $N(\theta, (nB_\theta)^{-1})$ where B_θ is the information matrix (section 2.12) for a single observation.

4.7 Restricted maximum-likelihood estimates

4.7.1 On certain occasions, when a family of distributions on a sample space is labelled by a vector-valued parameter θ, we have additional knowledge about the true parameter and we know that it satisfies certain restrictions. Then the parameter space Θ is expressed in the form

$$\Theta = \{\theta : \theta \in \mathbf{R}^s, h(\theta) = 0\},$$

where $h(\theta) = [h_1(\theta), h_2(\theta), \ldots, h_r(\theta)]$ is a vector-valued function mapping $\mathbf{R}^s$ into $\mathbf{R}^r$. Of course we wish an estimate of the true parameter to belong to Θ so that, as far as the method of maximum likelihood is concerned, we wish a restricted maximum-likelihood estimate, that is an estimate which maximizes the likelihood function subject to the restriction $h(\theta) = 0$.

4.7.2 As far as the theory of such restricted maximum-likelihood estimates is concerned, the natural mathematical approach is to reduce this case to that studied in section 4.6.1 by an initial re-parametrization. We 'fill out' the restricting functions $h_1, h_2, \ldots, h_r$ to a set $h_1, h_2, \ldots, h_r, h_{r+1}, \ldots, h_s$, in such a way that the function $h^* = (h_1, h_2, \ldots, h_s)$ is a one-to-one function from $\mathbf{R}^s$ onto itself. Then by setting $\phi_i = h_i(\theta_1, \theta_2, \ldots, \theta_s)$ $(i = 1, 2, \ldots, s)$

we obtain a new labelling of the family of possible distributions by the parameter $\phi = (0, 0, \ldots, 0, \phi_{r+1}, \ldots, \phi_s)$, whose first r components we may ignore, since they are all zero. Thus from a theoretical point of view this new problem is essentially the same as that of estimating an unrestricted parameter belonging to $\mathbf{R}^{s-r}$ and the properties established for the method of maximum likelihood in the latter case, asymptotic minimum-variance unbiasedness etc., will apply to restricted estimates also. Again only the bones of a rigorous argument are given here and these would require to be supplemented by regularity conditions and more details to complete the discussion.

4.7.3 The natural practical approach to the problem of finding restricted maximum likelihood estimates is a direct attack by the method of Lagrange multipliers, which leads to the restricted likelihood equations

$$D_\theta l(x, \theta) - H_\theta \lambda = 0$$
$$h(\theta) = 0,$$

where $\lambda = (\lambda_1, \lambda_2, \ldots, \lambda_r)'$ is a column-vector of Lagrange multipliers and H_θ is the $s \times r$ matrix of partial derivatives $\partial h_j(\theta)/\partial \theta_i$. With sufficient regularity, the restricted maximum-likelihood estimate $\hat{\theta}(x)$ emerges as a solution of these equations along with an appropriate Lagrange multiplier $\hat{\lambda}(x)$.

It is not possible to say much in general about this estimate $\hat{\theta}(x)$. However if we know that, with θ-probability near 1, $\hat{\theta}$ is very near θ, then the above restricted likelihood equations are approximately linear and by the same kind of argument as in section 4.6.1 we can obtain approximations to the distribution of $\hat{\theta}$. In particular, if we are dealing with a large sample then, subject to what is mild regularity from a practical point of view, it is true that $\hat{\theta}$ is very probably very near the true parameter θ – the argument of section 4.4 carries over with little modification, as the reader may verify. Let us suppose then that we are dealing with a sample of n, where n is large, so that $x = (x_1, x_2, \ldots, x_n)$ and the x_is are independent and identically distributed.

We have
$$D_\theta l(x, \hat{\theta}) - H_{\hat{\theta}} \hat{\lambda} = 0$$
$$h(\hat{\theta}) = 0,$$

and using Taylor's theorem to linearize about the true parameter θ (which we recall satisfies $h(\theta) = 0$), we have approximately

$$D_\theta l(x, \theta) + \{D_\theta^2 l(x, \theta)\} \{\hat{\theta} - \theta\} - H_{\hat{\theta}} \hat{\lambda} = 0$$
$$H'_\theta (\hat{\theta} - \theta) = 0.$$

The fact that the term $H_{\hat{\theta}} \hat{\lambda}$ can simply be replaced by $H_\theta \hat{\lambda}$ requires some explanation. This is because when $\hat{\theta}$ is near θ, it is also near the element $\hat{\theta}$ of $\mathbf{R}^s$ at which $l(x, \theta)$ takes its absolute maximum, so that $\hat{\lambda}$ is relatively small; hence when $H_{\hat{\theta}}$ is expanded about θ, the first-order terms in the expansion involve $\hat{\theta} - \theta$ and $\hat{\lambda}$ and so are of smaller order than those which we have included. Slight manipulation of these equations yields

$$\left[-\frac{1}{n} D^2 l(x, \theta)\right] \sqrt{n}(\hat{\theta} - \theta) + H_\theta \frac{1}{\sqrt{n}} \hat{\lambda} = \frac{1}{\sqrt{n}} D_\theta l(x, \theta)$$
$$H'_\theta \sqrt{n}(\hat{\theta} - \theta) = 0,$$

or, in matrix notation,

$$\begin{bmatrix} -\frac{1}{n} D^2 l(x, \theta) & H_\theta \\ H'_\theta & 0 \end{bmatrix} \begin{bmatrix} \sqrt{n}(\hat{\theta} - \theta) \\ \frac{1}{\sqrt{n}} \hat{\lambda} \end{bmatrix} = \begin{bmatrix} \frac{1}{\sqrt{n}} D_\theta l(x, \theta) \\ 0 \end{bmatrix}.$$

We now apply the law of large numbers to $-n^{-1} D^2 l(x, \theta)$ and find, as before, that this is approximately $\mathbf{B}_\theta$ (the information matrix for a single observation). The central limit theorem applied to $n^{-\frac{1}{2}} D_\theta l(x, \theta)$ shows that it is approximately an $N(0, \mathbf{B}_\theta)$ random variable. Carrying our approximation this one stage further shows, therefore, that approximately

$$\begin{bmatrix} \mathbf{B}_\theta & H_\theta \\ H'_\theta & 0 \end{bmatrix} \begin{bmatrix} \sqrt{n}(\hat{\theta} - \theta) \\ \frac{1}{\sqrt{n}} \hat{\lambda} \end{bmatrix} = \begin{bmatrix} Z \\ 0 \end{bmatrix}$$

where Z is $N(0, \mathbf{B}_\theta)$.

This is virtually the same set of equations as we had when dealing with the linear model subject to restrictions – which is not surprising since we arrived at this point by linearizing a non-linear set of equations. We can therefore carry over the results of section 3.10, simply by replacing $A'A$ by B_θ and H by H_θ and we find that, when B_θ has rank s,

$$\begin{bmatrix} \sqrt{n}(\hat{\theta}-\theta) \\ \dfrac{1}{\sqrt{n}}\lambda \end{bmatrix}$$

is approximately normally distributed with zero mean and variance matrix

$$\begin{bmatrix} P_\theta & 0 \\ 0 & -R_\theta \end{bmatrix}, \text{ where } \begin{bmatrix} B_\theta & H_\theta \\ H'_\theta & 0 \end{bmatrix}^{-1} = \begin{bmatrix} P_\theta & Q_\theta \\ Q'_\theta & R_\theta \end{bmatrix}.$$

4.7.4 *Non-identifiability and singularity of the information matrix*

There is a connexion between non-identifiability of a vector-valued parameter θ and singularity of the information matrix B_θ, which becomes clearer if we examine the function $z(\theta)$ introduced in section 4.4.

Suppose that a family of distributions is labelled by a parameter θ which ranges over an s-dimensional subset Θ of R^s, that is, Θ contains an s-dimensional rectangle. Let θ_0 be a particular element of Θ and θ a neighbouring point, and as in section 4.4.1,

let $z(\theta) = E_0\{\log p(x, \theta)\}$,

where $p(\cdot, \theta)$ is the density function on the sample space defining the distribution corresponding to the parameter θ. Let us suppose further that there is enough regularity in the family $\{p(\cdot, \theta) : \theta \in \Theta\}$ of density functions to permit the following operations, which we have already encountered.

$$\begin{aligned} z(\theta) &= E_0\{\log p(x, \theta)\} \\ &= E_0\{\log p(x, \theta_0) + [D_\theta \log p(x, \theta_0)]'(\theta-\theta_0) + \\ &\quad + (\theta-\theta_0)'[D_\theta^2 \log p(x, \theta_0)](\theta-\theta_0)\} + \text{terms of third order} \\ &= z(\theta_0) - (\theta-\theta_0)'M_{\theta_0}(\theta-\theta_0) + \text{small terms,} \end{aligned}$$

where M_θ is the information matrix for x. We know from our previous study of the function z, that if θ is identifiable, (that is, if different θs corresponding to different distributions) then $z(\theta_0)$ is an absolute maximum of z. This usually means in practice that the second-order terms in the expansion of $z(\theta)$ about $z(\theta_0)$ are negative, that is, that M_{θ_0} is positive definite. Of course this is not necessarily so. It is possible that $(\theta-\theta_0)'M_{\theta_0}(\theta-\theta_0)$ is zero and that higher order terms ensure that $z(\theta_0) > z(\theta)$. However this is unusual in practice, and usually identifiability of θ, together with the kind of regularity which permits the expansion of $z(\theta)$ indicated above, ensures that M_{θ_0} is positive definite, at least when θ_0 is an interior point of Θ.

Conversely, if M_{θ_0} is singular and so indefinite, this in practice usually means that there are parameters $\theta \neq \theta_0$ such that $z(\theta) = z(\theta_0)$ and this in turn means that there are different parameters yielding essentially the same distribution on the sample space, so that θ is not identifiable.

It is quite clear that any formal result connecting non-identifiability of θ and singularity of the information matrix which we might try to state would have to be hedged around by so many conditions that its content would be obscured. So we leave the discussion in this informal state, noting that usually lack of identifiability of θ implies singularity of the information matrix and vice versa.

4.7.5 For the linear model we discussed the possibility (section 3.4) of non-identifiability and the adjustment necessary to the technique for finding restricted estimates in this case. A similar adjustment is often possible in the above large-sample theory for restricted maximum-likelihood estimates in the case where the information matrix is singular. For, as indicated in section 4.7.4, this often means that θ is not identifiable without restrictions. However a number of the restrictions

$$h_i(\theta) = 0 \quad (i = 1, 2, \ldots, r),$$

say the first t of these, are just enough to ensure identifiability (as in the linear case); and this usually ensures that the matrix $B_\theta + H_{1\theta} H'_{1\theta}$ is positive definite, where $H_{1\theta}$ is the leading $s \times t$ submatrix of H_θ. The adjustment is now similar to that in the linear case. We replace B_θ by $B_\theta + H_{1\theta} H'_{1\theta}$ wherever it appears, and now $\sqrt{n}(\hat{\theta} - \theta)$ is approximately normally distributed with zero mean and variance matrix P_θ, the leading $s \times s$ submatrix in

$$\begin{bmatrix} B_\theta + H_{1\theta} H'_{1\theta} & H_\theta \\ H'_\theta & 0 \end{bmatrix}^{-1}.$$

For further details see Silvey (1959).

4.7.6 *Example*

In an experiment for measuring the DNA content of a particular type of cell, there is a chance of mistaking two cells for one, so that the experimental result may be a measurement of the DNA content of a single cell or of the sum of the contents of two cells. From the results $x = (x_1, x_2, \ldots, x_n)$ of a large number n of independent repetitions of this experiment, it is desired to estimate the mean and standard deviation of the DNA content of single cells.

In order that it should be possible to apply the method of maximum likelihood to this problem it is necessary to set up a model which involves only a finite number of unknown parameters. Now in this situation it is fairly realistic to assume that the DNA content of a single cell is normally distributed with unknown mean μ and unknown variance σ^2. There is an unknown probability α of mistaking two cells for one. If we further assume that when two cells are mistaken for one, these two cells may be regarded as independent of one another, then measurements resulting from the observation of two cells are normally distributed with mean 2μ and variance $2\sigma^2$. With these assumptions, the probability density on the line to describe the result of a single replicate of the experiment is

$$p^*(t, \theta) = (1-\alpha)\frac{1}{\sigma\sqrt{(2\pi)}}\exp\left[-\frac{(t-\mu)^2}{2\sigma^2}\right] + \alpha\frac{1}{\sqrt{2}\sigma\sqrt{(2\pi)}}\exp\left[-\frac{(t-2\mu)^2}{2\times 2\sigma^2}\right]$$

and the probability density on the sample space for n repetitions, R^n, is

$$p(x, \theta) = \prod_{i=1}^{n} p^*(x_i, \theta).$$

Here $\theta = (\alpha, \mu, \sigma)$ and we have a family of distributions on the sample space parametrized by a 3-vector, so that the method of maximum likelihood may be applied, in the same kind of way as in section 4.2.1.

Another way of setting up a model for this example is less sensible from a computational point of view but yields an illustration of the application of the method of computing restricted estimates. So we consider it for this reason. As before we assume that the DNA content of single cells is normally distributed, with unknown mean and variance which we now denote by μ_1 and σ_1^2 respectively. Again there is an unknown probability α of mistaking two cells for one. However we now assume that a measurement resulting from the observation of two cells is normally distributed with mean μ_2 and variance σ_2^2, so that the probability density on the line to describe the result of a single replicate of the experiment is now

$$p^*(t, \theta) = (1-\alpha)\frac{1}{\sigma_1\sqrt{(2\pi)}}\exp\left[-\frac{(t-\mu_1)^2}{2\sigma_1^2}\right] + \alpha\frac{1}{\sigma_2\sqrt{(2\pi)}}\exp\left[-\frac{(t-\mu_2)^2}{2\sigma_2^2}\right],$$

where $\theta = (\alpha, \mu_1, \sigma_1, \mu_2, \sigma_2)$, a 5-vector.

Correspondingly $p(x, \theta) = \prod_{i=1}^{n} p^*(x_i, \theta),$

and this density also involves five unknown parameters. If we are prepared to make the assumption that two cells mistaken for one are independent then $\mu_2 - 2\mu_1 = 0$
and $\sigma_2^2 - 2\sigma_1^2 = 0$,

and we may consider maximizing $p(x, \cdot)$ subject to these restrictions by the Lagrange multiplier technique of section 4.7.3. This of course is equivalent to the previous 'unrestricted maximization', but the reader may find it instructive to follow the theory for each case through in terms of this particular example.

Examples

4.1 Let $x_1, x_2, \ldots, x_n$ be a random sample from a distribution with density $p(x, \theta)$ depending on an unknown real parameter θ. Find the maximum-likelihood estimate of θ in the following cases.

(a) $p(\cdot, \theta)$ is the density function of a Poisson distribution with mean θ;
(b) $p(\cdot, \theta)$ is the density function of an exponential distribution,
$$p(x, \theta) = \theta e^{-x\theta} \quad (x > 0);$$
(c) $p(\cdot, \theta)$ is the density function of the uniform distribution on $(0, \theta)$.

In each case determine the distribution of the maximum-likelihood estimator. For cases (a) and (b) verify the large sample theory of chapter 4. Show that this theory is not applicable in case (c) and explain why.

4.2 On the Aegean island of Kalythos, the inhabitants suffer from a congenital eye disease whose effects become more marked with increasing age. Samples of fifty people were taken at five different ages and the numbers of blind people counted:

Age	20	35	45	55	70
Number of blind	6	17	26	37	44

It is conjectured that the probability of blindness at age x, $P(x)$, can be expressed in the form

$$P(x) = \{1 + e^{-(\alpha + \beta x)}\}^{-1}.$$

Comment on whether this hypothesis is reasonable, by constructing a suitable graph. Estimate α and β from the graph and then obtain maximum-likelihood estimates. Estimate also the age at which it is just more likely than not that an islander is blind.

4.3 A certain type of electrical component is manufactured in a large number of factories. The proportion p of defective components varies from factory to factory, and over factories p has approximately a B-distribution with density

$$\frac{p^{\alpha-1}(1-p)^{\beta-1}}{B(\alpha, \beta)},$$

where α and β are unknown parameters. Suppose that s factories are chosen at random and that n components produced by each are inspected. Given that m_i of the inspected components of the ith factory are defective ($i = 1, 2, \ldots, s$), explain in detail how to calculate maximum-likelihood estimates of α and β. Show that if $n = 1$, α and β are not identifiable.

4.4 Suppose that one has n pairs of measurements $(x_1, y_1), (x_2, y_2), \ldots, (x_n, y_n)$ the $2n$ values being distributed normally and independently with variance σ^2. The mean of x_i is ξ_i, that of y_i is η_i and the n pairs (ξ_i, η_i) lie on a circle centre (ξ, η) and radius ρ. It is required to estimate ξ, η and ρ. Obtain a maximum-likelihood solution of this problem, and elaborate the computational details. (*Camb. Dip.*)

4.5 In an experiment to measure the resistance of a crystal, independent pairs of observations $(x_i, y_i)(i = 1, 2, \ldots, n)$ of current x and voltage y are obtained. These are subject to errors (ε_i, η_i), so that

$$x_i = \mu_i + \varepsilon_i, \quad y_i = v_i + \eta_i,$$

where μ_i and v_i are the true values of current and voltage on the ith occasion and $v_i = \alpha \mu_i$, α being the resistance of the crystal.

On the assumption that the errors are independently and normally distributed with zero means and variances, var $\varepsilon_i = \sigma_1^2$, var $\eta_i = \sigma_2^2 = \lambda \sigma_1^2$, where λ is known, show that $\hat{\alpha}$, the maximum-likelihood estimator of α is a solution of the equation

$$\hat{\alpha}^2 S_{xy} + \hat{\alpha}(\lambda S_{xx} - S_{yy}) - \lambda S_{xy} = 0,$$

where $S_{xy} = \dfrac{1}{n}\sum x_i y_i, \quad S_{xx} = \dfrac{1}{n}\sum x_i^2, \quad S_{yy} = \dfrac{1}{n}\sum y_i^2.$

Show that, if $\sum \mu_i^2/n$ tends to a limit as $n \to \infty$, then $\hat{\alpha}$ is a consistent estimator of α.

Show that the method of maximum likelihood gives unsatisfactory results when λ is not assumed known. Explain why the standard theorems for maximum-likelihood estimators do not apply to this problem. (*Camb. Dip.*)

4.6 A radioactive sample emits particles randomly at a rate which decays with time, the rate being $\lambda e^{-\kappa t}$ after time t. The first n particles emitted are observed at successive times $t_1, t_2, \ldots, t_n$. Set up equations for maximum-likelihood estimates $\hat{\lambda}$ and $\hat{\kappa}$, and show that $\hat{\kappa}$ satisfies the equation

$$\frac{\hat{\kappa} t_n}{e^{\hat{\kappa} t_n} - 1} = 1 - \hat{\kappa} \bar{t},$$

where $\bar{t} = \dfrac{1}{n} \sum_{i=1}^{n} t_i.$

Find a simple approximation for $\hat{\kappa}$ when $\hat{\kappa} t_n$ is small. (*Camb. Dip.*)

4.7 A cell contains granules which may be regarded as spheres of equal but unknown radius r, and which may be assumed to be distributed randomly throughout the cell. In order to estimate r, a section of the cell is observed under a microscope and this section contains circular sections of n granules. If the radii of these sections are $x_1, x_2, \ldots, x_n$, determine the maximum-likelihood estimate of r. What is its distribution, for large n?

4.8 Let $x_1, x_2, \ldots, x_n$ be a random sample from the exponential distribution with density $\theta_1 e^{-\theta_1 x} (x > 0)$ and $y_1, y_2, \ldots, y_n$ an independent random sample from a distribution with density $\theta_2 e^{-\theta_2 y}$ $(y > 0)$. Find maximum-likelihood estimates of θ_1 and θ_2. Find, directly, restricted maximum-likelihood estimates subject to the condition $\theta_1 = \theta_2$ and verify the general theory of restricted estimates for this case.

5 Confidence Sets

The method of maximum likelihood is appealing because, as we have said, in some sense a maximum-likelihood estimate is the most plausible parameter value after an observation x has been made – it is that value of the parameter which gives greatest probability to x. However while we may accept this, we should be extremely reluctant to believe that a maximum-likelihood estimate coincided with the true parameter in all circumstances, and it is natural to ask how near the true parameter we might expect a maximum-likelihood estimate (or indeed any estimate) to be. The very use of the phrase 'how near' implies that there is a metric on the parameter space, but it is useful to think about this question in more general terms.

We may take the point of view that, when an observation has been made, this observation divides the parameter set into two disjoint subsets: a 'plausible' subset and an 'implausible' subset; and that what we really want to do, rather than to fix attention on a particular parameter value as an estimate of the true parameter, is to determine this plausible subset of parameter values. Then our conclusion based on an observation would be, 'The true parameter is in such-and-such a subset of the set of possible parameters.' The formalization of this idea leads to the problem of *set estimation*.

5.1 Confidence interval

It is possibly helpful to initiate this discussion by considering a particular example. Suppose that we have available a random sample $x = (x_1, x_2, \ldots, x_n)$ from a normal distribution with unknown mean μ and unknown variance σ^2 and we wish to determine a 'plausible' set of values of μ. We may argue as follows.

Let $\theta = (\mu, \sigma^2)$ and let $t(x, \mu) = \sqrt{n}(\bar{x} - \mu)/s$, where $\bar{x} = n^{-1} \sum x_i$ and $s^2 = (n-1)^{-1} \sum (x_i - \bar{x})^2$. The θ-distribution of $t(x, \mu)$ is known; it is distributed as Student's t with $n-1$ degrees of freedom, in statistical jargon. Hence, *without knowing what θ is*, we can find a number t_α such that

$$P_\theta\{-t_\alpha \leqslant t(x, \mu) \leqslant t_\alpha\} = 1 - \alpha,$$

where α is a small preassigned number between 0 and 1. This may be rewritten in the following form:

$$P_\theta\left[\bar{x}-t_\alpha\frac{s}{\sqrt{n}} \leqslant \mu \leqslant \bar{x}+t_\alpha\frac{s}{\sqrt{n}}\right] = 1-\alpha,$$

and read as: 'Whatever the true value of μ may be, the probability that the random interval

$$\left[\bar{x}-t_\alpha\frac{s}{\sqrt{n}}, \bar{x}+t_\alpha\frac{s}{\sqrt{n}}\right]$$

contains this true value is $1-\alpha$.' Thus for each given x, the interval

$$\left[\bar{x}-t_\alpha\frac{s}{\sqrt{n}}, \bar{x}+t_\alpha\frac{s}{\sqrt{n}}\right]$$

may be regarded as a 'plausible' set of values of μ, plausible in the sense that we are $(100(1-\alpha))$ per cent confident that this set contains the true parameter value. The interval is called a *confidence interval* for μ with *confidence coefficient* $1-\alpha$.

Note that as α decreases, t_α increases, so that this $100(1-\alpha)$ per cent confidence interval, corresponding to any given x, widens as α decreases. In other words, if we wish to have great confidence in a chosen plausible interval, we must choose a larger interval than is necessary if we are content to have less confidence in our chosen interval; and this is not surprising.

Note also that for fixed α and any given x, there is not a *unique* $100(1-\alpha)$ per cent confidence interval for μ. There is no reason why we should not choose two numbers $t_{1\alpha}$ and $t_{2\alpha}$ such that $t_{1\alpha} \neq -t_{2\alpha}$, and still have

$$P_\theta\{t_{1\alpha} \leqslant t(x,\mu) \leqslant t_{2\alpha}\} = 1-\alpha.$$

If we do so, then we are led by exactly the same argument to the $100(1-\alpha)$ per cent confidence interval

$$\left[\bar{x}-t_{2\alpha}\frac{s}{\sqrt{n}}, \bar{x}-t_{1\alpha}\frac{s}{\sqrt{n}}\right].$$

Having observed an x, we may equally well say that we are $100(1-\alpha)$ per cent confident that the true value of μ lies in this interval.

5.2 General definition of a confidence set

This example illustrates one interpretation of what we mean by dividing a parameter set into a subset of plausible values and a subset of implausible values after an observation has been taken. It may not be a completely convincing interpretation, and indeed it is the subject of some criticism to which we shall return later when discussing what is known as Bayesian inference. However, since it has considerable practical value at least, we shall now establish this idea in a general setting.

Our basic mathematical framework is as before – a sample space X, a family $\{P_\theta\}$ of probability distributions on X, this family being labelled by the parameter θ which ranges over a parameter space Θ. Now suppose that $\{S_x; x \in X\}$ is a family of subsets of Θ with the property that, for all θ,

$$P_\theta\{x : S_x \supset \theta\} = 1 - \alpha. \tag{5.1}$$

The set S_x is then called a *confidence set for* θ with confidence coefficient $1 - \alpha$. The practical interpretation of this is that if we observe the point x in the sample space and conclude that the true parameter belongs to the subset S_x of the parameter space, we can have $100(1-\alpha)$ per cent confidence in this conclusion.

Similarly, if $\{S_x; x \in X\}$ is a family of subsets of Θ with the property that, for all θ,

$$P_\theta\{x : S_x \supset \theta\} \geq 1 - \alpha,$$

then S_x is a confidence set for θ with *lower* confidence coefficient $1 - \alpha$. Having observed x we can then be *at least* $100(1-\alpha)$ per cent confident that the true parameter is in S_x.

5.3 Construction of confidence sets

5.3.1 It is one thing to define a family $\{S_x; x \in X\}$ of subsets with the property **5.1** and another to construct such a family for any given problem. In section 5.1 we illustrated one way of doing this in the case of a real parameter, and the general statement of this method is as follows. If we can find a function $t(x, \theta)$, called a pivotal quantity, whose θ-distribution does not depend on θ, and which, for each fixed x, is a monotonic function of θ (θ being a real parameter), then we can apply the method of section 5.1. For then we can find t_1 and t_2 such that

$$P_\theta\{t_1 < t(x, \theta) < t_2\} = 1 - \alpha,$$

and, $t(x, \cdot)$ being monotonic, we may rewrite the inequality in braces in the form

$$\theta_1(x) < \theta < \theta_2(x),$$

from which we derive $[\theta_1(x), \theta_2(x)]$ as a confidence interval for θ. While this method is applicable in certain important practical problems, it is scarcely of sufficient generality to warrant trying to find necessary and sufficient conditions for its applicability. However, it is often relatively easy to find an appropriate pivotal quantity when normality of the family of distributions on the sample space may be assumed. The example of section 5.1 is one illustration. As another, we consider linear regression with one concomitant variable, that is, we have the model

$$x_i = \alpha + \beta a_i + \varepsilon_i \quad (i = 1, 2, \ldots, n),$$

where α and β are unknown parameters, the a_is are known values of a concomitant variable and the ε_is are independent normally distributed errors each with mean zero and unknown variance σ^2. Suppose that we wish to determine a confidence interval for β.

The least-squares estimate of β (it is also the maximum-likelihood estimate when, as here, normality of errors is assumed) is

$$\hat{\beta} = \frac{\sum (a_i - \bar{a}) x_i}{\sum (a_i - \bar{a})^2},$$

where of course $n\bar{a} = \sum_{i=1}^{n} a_i$; and $\hat{\beta}$ is normally distributed with mean β and variance $\sigma^2 / \sum (a_i - \bar{a})^2$. Since σ^2 is unknown, it is natural to consider replacing it by the unbiased estimate

$$s^2 = \frac{1}{n-2} \sum (x_i - \hat{\alpha} - \hat{\beta} a_i)^2,$$

the 'residual mean square', and to try as a pivotal quantity

$$\frac{(\hat{\beta} - \beta)\sqrt{\sum (a_i - \bar{a})^2}}{s}.$$

Since the distribution of this function does not depend on anything unknown – it is distributed as Student's t with $n-2$ degrees of freedom – it is indeed a pivotal quantity, and we may use it to construct the $100(1-\alpha)$ per cent confidence interval for β,

$$\left[\hat{\beta} - t_\alpha \frac{s}{\sqrt{\sum(a_i - \bar{a})^2}},\ \hat{\beta} + t_\alpha \frac{s}{\sqrt{\sum(a_i - \bar{a})^2}} \right].$$

A similar argument may be used to construct a confidence interval for α, and for any linear combination $\alpha + c\beta$. Consider the latter: $\hat{\alpha} + c\hat{\beta}$ is normally distributed with mean $\alpha + c\beta$ and variance

$$\sigma^2 \left[\frac{1}{n} + \frac{(\bar{a} - c)^2}{\sum (a_i - \bar{a})^2} \right].$$

As above

$$\frac{\hat{\alpha} + c\hat{\beta} - (\alpha + c\beta)}{s\sqrt{\left[\dfrac{1}{n} + \dfrac{(\bar{a} - c)^2}{\sum (a_i - \bar{a})^2}\right]}}$$

is a pivotal quantity, being distributed as Student's t with $n-2$ degrees of freedom. Hence the interval

$$\left[\hat{\alpha} + c\hat{\beta} - t_\alpha s \sqrt{\left(\frac{1}{n} + \frac{(\bar{a} - c)^2}{\sum (a_i - \bar{a})^2}\right)},\ \hat{\alpha} + c\hat{\beta} + t_\alpha s \sqrt{\left(\frac{1}{n} + \frac{(\bar{a} - c)^2}{\sum (a_i - \bar{a})^2}\right)} \right]$$

is a $100(1-\alpha)$ per cent confidence interval for $\alpha + c\beta$.

5.3.2 Another important practical application of confidence sets occurs when the family of possible distributions on the sample space is parametrized by a vector-valued parameter θ and we are dealing with a large sample. Then although we do not necessarily assume normality of the family $\{P_\theta\}$, large-sample theory enables us to assume approximate normality of the maximum-likelihood estimator, $\hat\theta$. Subject to regularity, $\hat\theta$ is approximately normal with mean θ and variance matrix $V = (n B_{\hat\theta})^{-1}$, where B_θ is the information matrix for a single observation. It follows that

$$(\hat\theta-\theta)' V^{-1}(\hat\theta-\theta)$$

is distributed approximately as χ^2 with s degrees of freedom (θ being an s-vector). Consequently we can find a number χ_α^2 such that, for all θ,

$$P_\theta\{(\hat\theta-\theta)' V^{-1}(\hat\theta-\theta) \leq \chi_\alpha^2\} = 1-\alpha.$$

For a given x, then, the set of θ satisfying

$$[\theta-\hat\theta(x)]' V^{-1}[\theta-\hat\theta(x)] \leq \chi_\alpha^2$$

is a $100(1-\alpha)$ per cent confidence set for θ. This set is an s-dimensional ellipsoid centred on $\hat\theta(x)$.

5.3.3 *Example*

Consider again the linear regression model of section 5.3.1, assuming now that the errors are independent and identically distributed, but not assuming normality of these errors. Suppose that n is large and that we wish to determine a confidence set for $\theta = (\theta_1, \theta_2)$, where $\theta_1 = \alpha + c_1 \beta$ and $\theta_2 = \alpha + c_2 \beta$. For reasonably spread-out values $a_1, a_2, \ldots, a_n$ of the concomitant variable, the least-squares estimators $\hat\alpha$ and $\hat\beta$ are approximately jointly normally distributed. This may be proved by a multivariate central limit theorem (see Cramér, 1937, p. 113). Moreover, by least-squares theory, $\hat\alpha$ and $\hat\beta$ have respective means α and β, and variance matrix

$$\sigma^2 \begin{bmatrix} n & \sum a_i \\ \sum a_i & \sum a_i^2 \end{bmatrix}^{-1}.$$

Hence $\hat\theta_1 = \hat\alpha + c_1 \hat\beta$ and $\hat\theta_2 = \hat\alpha + c_2 \hat\beta$ are approximately jointly normal with means θ_1 and θ_2 respectively and variance matrix

$$V_\sigma = \sigma^2 \begin{bmatrix} \dfrac{1}{n} + \dfrac{(c_1-\bar a)^2}{\sum(a_i-\bar a)^2} & \dfrac{1}{n} + \dfrac{(c_1-\bar a)(c_2-\bar a)}{\sum(a_i-\bar a)^2} \\ \dfrac{1}{n} + \dfrac{(c_1-\bar a)(c_2-\bar a)}{\sum(a_i-\bar a)^2} & \dfrac{1}{n} + \dfrac{(c_2-\bar a)^2}{\sum(a_i-\bar a)^2} \end{bmatrix}.$$

If n is large and $\sum(a_i-\bar a)^2 = O(n)$, then to order n^{-1}, $V_\sigma = V_s$ where s^2, the residual mean square, is an unbiased estimate of σ^2. Consequently $\hat\theta = (\hat\theta_1, \hat\theta_2)$ is approximately normal with mean (θ_1, θ_2) and variance matrix V_s, and it follows that $(\hat{\theta}-\theta)'V_s^{-1}(\hat{\theta}-\theta)$ is distributed approximately as χ^2 with two degrees of freedom. We can now use the argument of section 5.3.2 to obtain an elliptic confidence set for θ.

This example is not of course a direct application of the argument of section 5.3.2 in that we are not using the method of maximum likelihood. However it is in exactly the same spirit and it illustrates the kind of approximations made in large sample maximum-likelihood theory. It may seem that there is a great deal of approximation going on and that the resulting confidence set is only very approximately a $100(1-\alpha)$ per cent confidence set. This may well be true, but it does not give cause for great concern in practice if we adopt the point of view that we are using observations to divide possible parameters into a plausible set and implausible set. Then it is adequate to know roughly what we mean by the plausible set.

5.4 Optimal confidence sets

As has been indicated already, there is nothing unique about a confidence set of given confidence level. This raises the question of which of many $100(1-\alpha)$ per cent confidence sets we should choose in a given situation. It may be that we want one which is in some sense 'smallest' and this can be interpreted as one whose probability of covering false values of the parameter is uniformly minimum (if such a confidence set exists). It may be that extraneous aspects of the problem at hand require a confidence set of given 'shape'. Indeed there is a considerable body of theory concerned with the problems of the existence and construction of a confidence set which is optimum relative to some stated criteria, and much of this is closely connected with the theory of optimum tests of hypotheses which we shall discuss in a subsequent chapter. The reader, however, may find that the confidence set interpretation of the basic intuitive idea of a subset of parameter values becoming relatively plausible after an observation has been made is less convincing than the Bayesian interpretation which again we shall discuss later. So we shall not pursue the problem of optimum confidence sets further at this point.

Examples

5.1 Let $x_1, x_2, \ldots, x_n$ be a random sample from a $N(\mu, \sigma^2)$ distribution with μ and σ^2 unknown, and let

$$s^2 = \frac{1}{n-1} \sum (x_i - \bar{x})^2.$$

Show that s^2/σ^2 is a pivotal quantity which may be used to construct a confidence interval for σ^2.

5.2 Given a random sample $x_1, x_2, \ldots, x_n$ from the exponential distribution with density $\theta e^{-x\theta}$, construct a 95 per cent confidence interval for θ.

5.3 Let x be $N(\mu, \sigma^2)$ with μ and σ unknown. Show how $(x-\mu)/\sigma$ may be used to construct a wedge-shaped confidence set for $\theta = (\mu, \sigma)$.

5.4 If $x_1, x_2, \ldots, x_n$ is a random sample from an $N(\mu, \sigma^2)$ distribution, then $(\bar{x}-\mu)/s$ may be used to find a $100(1-\alpha)$ per cent confidence interval for μ, $(\underline{\mu}, \bar{\mu})$, say; while s^2/σ^2 may be used to find a $100(1-\alpha)$ per cent confidence interval for σ, $(\underline{\sigma}, \bar{\sigma})$. Here $s^2 = \sum(x_i - \bar{x})^2/(n-1)$. Consider the rectangular region in the (μ, σ) plane $\{(\mu, \sigma): \underline{\mu} \leq \mu \leq \bar{\mu}, \underline{\sigma} \leq \sigma \leq \bar{\sigma}\}$. This may be taken as a confidence region for (μ, σ). What can be said about its confidence coefficient?

5.5 Let x and y be independent random variables with densities $\lambda e^{-\lambda x}(x > 0)$ and $\mu e^{-\mu y}(y > 0)$, respectively. Show that

$$C_{x,y} = \{(\lambda, \mu): \lambda x + \mu y \leq a\}$$

is a confidence region for (λ, μ) with confidence coefficient $1-(1+a)e^{-a}$.

5.6 If x is an unbiased estimator of ξ with known variance σ_1^2 and y is an unbiased estimator of η with known variance σ_2^2 and if x and y are independent and such that, for any λ, $x - \lambda y$ may be taken to be normally distributed, show how to obtain 95 per cent confidence limits for the ratio ξ/η.

These limits will sometimes include the whole real axis, and it has been suggested that such cases should be omitted from the calculation of the proportion P of times that the confidence interval covers the true value. Show that, if this were done, the proportion P would depend on ξ/η.

6 Hypothesis Testing

What is sometimes termed the classical theory of statistical inference – though it is not more than forty years old – has two main branches, the theories of estimation and hypothesis testing. We have been concerned with the former in previous chapters and we now consider some of the main aspects of the latter.

The basic mathematical framework is as before: a sample space X and a family $\{P_\theta : \theta \in \Theta\}$ of probability distributions on X, labelled by a parameter θ which ranges over a parameter space Θ. Points in X are the mathematical representation of possible observations; the family $\{P_\theta\}$ represents possible descriptions of the inherent variability in the observational situation being considered; and one member of this family is the true description, though which one is unknown.

When we are dealing with a real situation in which observations may be made and which are described by a probabilistic model, a scientific hypothesis is a statement regarding the probabilistic structure describing the inherent variability in the observational situation. For example, suppose that a very large population is classified according to two factors A and B, that there are r different categories $A_1, A_2, \ldots, A_r$ of the factor A and s categories $B_1, B_2, \ldots, B_s$ of B. Each individual in the population belongs to one and only one of the rs cells $A_i B_j$, and the proportion θ_{ij} of the population in the cell $A_i B_j$ is unknown, $i = 1, 2, \ldots, r, j = 1, 2, \ldots, s$. An individual chosen at random from this population then has probability θ_{ij} of falling in the cell $A_i B_j$. If we observe the numbers in a random sample of n individuals belonging to the different cells, then a typical observation x takes the form $x = (n_{11}, n_{12}, \ldots, n_{rs})$ where n_{ij} is the number of individuals in the cell $A_i B_j$; and the appropriate family of possible distributions on the sample space is the family of multinomial distributions, parametrized by $\theta = (\theta_{11}, \theta_{12}, \ldots, \theta_{rs})$. The parameter space Θ is $\{\theta : 0 \leqslant \theta_{ij} \leqslant 1, \sum_{i,j} \theta_{ij} = 1\}$.

Consider the hypothesis that 'there is no association between the factors A and B'. This hypothesis is translated into the language of our multinomial model as follows:

Let $\theta_{i.} = \sum_{j=1}^{s} \theta_{ij}$ and $\theta_{.j} = \sum_{i=1}^{r} \theta_{ij}$.

Then the hypothesis states that for all i and j

$$\theta_{ij} = \theta_{i.}\theta_{.j}.$$

Thus the hypothesis imposes a further limitation on the class of possible distributions. Under it the family of possible distributions is the multinomial family parametrized by θ, and the set of possible θ is

$$\{\theta : \theta \leqslant \theta_{ij} \leqslant 1, \sum_{i,j} \theta_{ij} = 1, \theta_{ij} = \theta_{i.}\theta_{.j}; \text{ for all } i \text{ and } j\},$$

a subset of the parameter space Θ.

Generally this is the case. *A hypothesis is a statement which implies that the true probability distribution describing the inherent variability in an observational situation belongs to a proper subset of the family of possible probability distributions.* Alternatively we may say that a hypothesis implies that the true parameter θ belongs to a proper subset of the parameter space Θ; and it is convenient to identify the hypothesis with the subset, to talk about the hypothesis ω, where $\omega \subset \Theta$.

The theory of hypothesis testing is concerned with the problem: 'Is a given observation consistent with some stated hypothesis or is it not?' A statistical test of a hypothesis is a rule which assigns each possible observation to one of two exclusive categories: 'consistent with the hypothesis under consideration' and 'not consistent with this hypothesis'.

Thus in terms of our mathematical model, a hypothesis defines a subset ω of the parameter space Θ; a statistical test partitions the sample space X into two subsets, a set of points each of which is consistent with ω; and its complement, consisting of points not consistent with ω. There are then as many tests of a given hypothesis ω as there are subsets of X. Our problem is to choose one which is 'good' in some sense.

In simple situations, there is often a relatively small class of tests which seem worth considering on a purely intuitive basis. For instance suppose that a new drug is being considered with a view to curing a certain disease. The drug is given to n patients suffering from the disease and the number r of cures is noted. We wish to test the hypothesis that there is at least a 50–50 chance of a cure by this drug. Here our sample space X is simple – it is the set $\{0, 1, 2, \ldots, n\}$. The family $\{P_\theta\}$ of possible distributions on X is (assuming independent patients) the family of binomial distributions, parametrized by the real parameter θ taking values in $[0, 1]$ – θ being interpreted as the probability of cure. The stated hypothesis defines the subset $\omega = [\frac{1}{2}, 1]$ of the parameter space. And the only tests of ω which seem worth considering at all are those for which the set of x taken to be consistent with ω have the form $\{x : x \geqslant k\}$. On the face of it, it would seem absurd to consider that r cures out of n patients were consistent with ω, while $r+1$ were not. Indeed we may be tempted to go further and say that there is only one 'reasonable' test for this hypothesis ω, namely that for which the set of x taken to be consistent with ω is $\{x : x \geqslant \frac{1}{2}n\}$; though we must bear in mind that less than $\frac{1}{2}n$ cures are possible even if the probability θ of cure is greater than a half.

In order to build up a theory of hypothesis testing and establish methods which can be applied in situations which are too complicated to be dealt with on an intuitive basis, it is necessary to analyse the intuitive reasoning which leads to 'reasonable' rests for simple situations. The key ideas of the 'classical' analysis were provided by Neyman and Pearson.

6.1 The Neyman–Pearson theory

We first change our terminology to agree with that of the classical theory. According to this, there are in a hypothesis-testing problem, two hypotheses involved, the hypothesis ω of primary interest and the complementary hypothesis $\Theta - \omega$. The first of these is called the null-hypothesis and the second the alternative hypothesis; and the mere fact that they are given different names suggests that in some sense they are not on an equal footing, a point to which we shall return. A statistical test of ω against the alternative $\Theta - \omega$ partitions the sample space into a region of acceptance of ω – what we have called the set consistent with ω – and its complementary region, a region of rejection of ω (acceptance of $\Theta - \omega$), called the *critical region* of the test of ω against $\Theta - \omega$. Our object in constructing a 'good' test may then be interpreted as choosing a critical region which is optimum relative to some criterion.

The Neyman–Pearson criterion is based on recognition of the fact that with any statistical test are associated two possible errors.

(a) We may reject ω when it is true, that is, when the true parameter belongs to ω. This is called a Type I error.

(b) We may accept ω when it is false, that is, when the true parameter belongs to $\Theta - \omega$. This is called a Type II error.

Associated with any test, then, are two functions which describe the probabilities of error characteristic of this test, and these functions are as follows.

Let the critical region of the test be R; a subset of the sample space. The function α is defined on ω by

$$\alpha(\theta) = P_\theta(R)$$

and this function describes the probabilities of the Type I error. Similarly the function β, defined on $\Theta - \omega$ by

$$\beta(\theta) = P_\theta(\bar{R}) = 1 - P_\theta(R)$$

describes the probabilities of the Type II error. The function $1 - \beta(\theta)$ is called the *power function* of the test.

A test whose error probabilities are as small as possible is clearly desirable. However, equally clearly, we cannot choose R in such a way that $\alpha(\theta)$ and $\beta(\theta)$ are simultaneously uniformly minimized, except in very special circumstances. By taking $R = \emptyset$, the empty set, we can make $\alpha(\theta) \equiv 0$ and by taking $R = X$, we can make $\beta(\theta) \equiv 0$. Hence a test which uniformly minimized both error-probability functions would require to have zero error probabilities, and usually no such test exists. The position is analogous to that in estimation which we have already encountered. To demand of a test that it uniformly minimizes error probabilities is analogous to the demand that an estimator of a real parameter have uniformly minimum mean-square error. Normally we can achieve neither. So again we must modify our demands.

The modification suggested by Neyman and Pearson is based on the fact that in most circumstances our attitudes to the hypotheses ω and $\Theta - \omega$ are different – we are often asking if there is sufficient evidence to reject the hypothesis ω. In terms of the two possible errors this may be translated into the statement that often the Type I error is more serious than the Type II error. Consequently we should control the probability of the Type I error at some pre-assigned small value α, and then, subject to this control, look for a test which uniformly minimizes the function describing the probabilities of Type II error. In other words we should limit consideration to tests which satisfy the condition

$$\alpha(\theta) \leqslant \alpha \quad \text{for all } \theta \in \omega,$$

and among these choose that one, if it exists, for which $\beta(\theta)$ is uniformly minimized on $\Theta - \omega$; or, equivalently, for which $1 - \beta(\theta)$ is uniformly maximized on $\Theta - \omega$.

Now a test in a given class whose power function is uniformly no larger than that of any other test in the class is said to be *uniformly most powerful*.

If a test satisfies the condition

$$\alpha(\theta) \leqslant \alpha \quad \text{for all } \theta \in \omega,$$

α is called *the significance level* of the test. (Incidentally

$$\sup_{\theta \in \omega} \alpha(\theta)$$

is called the *size* of the test). Hence the Neyman–Pearson theory may be summarized by the statement:

An optimum test is a uniformly-most-powerful (U.M.P.) test of given significance level α.

6.2 Simple hypotheses

While the criterion of optimality just stated may not be universally acceptable, it is appropriate enough in certain circumstances to warrant further investigation. Immediately we are faced again with the problem of the existence of such a U.M.P. test, a problem which we shall now consider.

We start from the simplest possible situation, that where Θ has only two elements θ_0 and θ_1, say, and where $\omega = \{\theta_0\}$, $\Theta - \omega = \{\theta_1\}$. A hypothesis which specifies a set in the parameter space containing only one element is called a *simple* hypothesis. Thus in statistical terminology we are now considering testing a simple null-hypothesis against a simple alternative, a problem of limited practical interest but of considerable analytic content. In this case, the power function of any test reduces to a single number, so that the 'uniformly' in U.M.P. becomes redundant, and we examine the question of the existence of a most-powerful test of given significance level α.

Suppose that the probability distributions P_{θ_0} and P_{θ_1} on the sample space X are defined by density functions p_0 and p_1 respectively with respect to some fixed measure on X. (There is no loss of generality in this assumption since the fixed measure may be taken, for instance, to be $P_{\theta_0} + P_{\theta_1}$. We shall denote by dx an element of the fixed measure so that $P_{\theta_i}(E) = \int_E p_i(x)\,dx$; when X is Euclidean and the fixed measure is Lebesgue measure this is a natural notation; if X is discrete and the fixed measure is natural counting measure, $\int_E p_i(x)\,dx$ is to be interpreted as $\sum_{x_j \in E} p_i(x_j)$.) The key result on most-powerful tests is the Neyman–Pearson fundamental lemma, of which we now prove a limited version.

6.2.1 Fundamental lemma

Let R be any region of the sample space such that $P_{\theta_0}(R) \leq \alpha$. Suppose that there exists a region R^* of X of the form $R^* = \{x : p_1(x)/p_0(x) \geq k\}$ and such that $P_{\theta_0}(R^*) = \alpha$. Then $P_{\theta_1}(R^*) \geq P_{\theta_1}(R)$.

Proof. We shall comment later on the existence of an R^* satisfying the stated conditions. Assume for the moment that such an R^* does exist. Denote by $\bar{R}$ and $\bar{R}^*$ the complements of R and R^* respectively.

Then $P_{\theta_1}(R^*) - P_{\theta_1}(R) = \int_{R^* \cap \bar{R}} p_1(x)\,dx - \int_{\bar{R}^* \cap R} p_1(x)\,dx.$

Now on $R^* \cap \bar{R}$ $p_1(x) \geq k p_0(x)$,

and so $\int_{R^* \cap \bar{R}} p_1(x)\,dx \geq k \int_{R^* \cap \bar{R}} p_0(x)\,dx.$

Similarly $\int_{\bar{R}^* \cap R} p_1(x)\,dx < k \int_{\bar{R}^* \cap R} p_0(x)\,dx.$

Therefore $P_{\theta_1}(R^*) - P_{\theta_1}(R) \geq k \left[\int_{R^* \cap \bar{R}} p_0(x)\,dx - \int_{\bar{R}^* \cap R} p_0(x)\,dx \right]$

$= k \left[\int_{R^*} p_0(x)\,dx - \int_R p_0(x)\,dx \right]$

$= k [P_{\theta_0}(R^*) - P_{\theta_0}(R)]$

$\geq 0,$

since $k \geq 0$, $P_{\theta_0}(R^*) = \alpha$ and $P_{\theta_0}(R) \leq \alpha$;

i.e. $P_{\theta_1}(R^*) \geq P_{\theta_1}(R)$,

which completes the proof.

This result, that R* is the critical region of a most-powerful test of significance level α of $\{\theta_0\}$ against $\{\theta_1\}$, is one of considerable intuitive appeal. Suppose that we set out to order points in the sample space according to the amount of evidence they provide for θ_1 rather than θ_0. We should naturally order them according to the value of the ratio $p_1(x)/p_0(x)$; any x for which this ratio is large provides evidence that θ_1 rather than θ_0 is the true parameter. And if we must choose a subset of possible observations which indicate that θ_1 is the true parameter, then it seems sensible to put into this subset those xs for which the ratio $p_1(x)/p_0(x)$ is large – in other words to choose a subset of the form $\{x : p_1(x)/p_0(x) \geq k\}$. The Neyman–Pearson analysis, based on probabilities of error, now gives us a basis for choosing k; we should choose k, so that, if possible,

$$P_{\theta_0}\left\{x : \frac{p_1(x)}{p_0(x)} \geq k\right\} = \alpha.$$

If P_{θ_0} and P_{θ_1} are discrete, it will be possible to satisfy this equation only for very special values of α. For example, suppose that an observation x is the number of successes in three independent trials; the probability of success is either $\theta_0 = \frac{1}{4}$ or $\theta_1 = \frac{3}{4}$; and we wish to test the null hypothesis $\{\theta_0\}$ against the alternative $\{\theta_1\}$ at significance level $\alpha = 0.05$. It is fairly obvious, and easily verified, that the ratio $p_1(x)/p_0(x)$ is an increasing function of x. Consequently the problem of finding k such that

$$P_{\theta_0}\left\{x : \frac{p_1(x)}{p_0(x)} \geq k\right\} = \alpha = 0.05$$

is equivalent to the problem of finding k' such that

$$P_{\theta_0}\{x : x \geq k'\} = 0.05.$$

Now x can take only the values 0, 1, 2, 3; and if $2 < k' \leq 3$,

$$P_{\theta_0}\{x : x \geq k'\} = P_{\theta_0}\{x = 3\} = \tfrac{1}{64} < 0.05,$$

while if $1 < k' \leq 2$,

$$P_{\theta_0}\{x : x \geq k'\} = P_{\theta_0}\{x = 2 \text{ or } x = 3\} = \tfrac{10}{64} > 0.05.$$

It now becomes clear that there is no k' such that

$$P_{\theta_0}\{x : x \geq k'\} = 0.05$$

and consequently no k such that

$$P_{\theta_0}\left\{x : \frac{p_1(x)}{p_0(x)} \geq k\right\} = 0.05.$$

Referring then to the fundamental lemma, this is a situation where there does not exist a set R* satisfying the conditions stated in that lemma. Hence we cannot use the lemma to construct a most-powerful test of $\{\theta_0\}$ against $\{\theta_1\}$ of significance level α.

6.2.2 There are two ways of overcoming this difficulty, one practical, the other more designed for mathematical elegance. The practical outlook is as follows.

The test to which we are led by the fundamental lemma is a *likelihood-ratio test* of *size* α, and the lemma may be regarded as providing support for likelihood-ratio tests. The fact that in particular cases there does not exist one whose size is exactly equal to α does not matter; we merely use that likelihood-ratio test of significance level α, whose size is as nearly as possible α. Thus in the discrete example which we have just been discussing, this outlook would lead us to use the test whose critical region consisted of the single element $x = 3$, a test of size $\frac{1}{64}$. According to this, only the observation $x = 3$ would provide sufficient evidence for rejecting the null-hypothesis that the probability of success was one quarter.

6.2.3 The 'mathematical' method of overcoming the difficulty caused by possible discreteness of the probability distributions involved is to allow 'randomized' tests, according to which, having observed an x in the sample space, with probability $\phi(x)$ we decide that the alternative hypothesis is true and with probability $1 - \phi(x)$ we decide that the null-hypothesis is true. Thus any non-negative function ϕ on the sample space taking values between 0 and 1 defines a statistical test, and in particular those functions which take only the values 0 and 1 define the type of tests which we have been considering up to this point, namely non-randomized tests. If we adopt this more general view of what constitutes a statistical test, then we may refer unambiguously to the test ϕ, where ϕ is a function on the sample space such that $0 \leqslant \phi(x) \leqslant 1$, for all x. Then for testing a hypothesis ω against the alternative $\Theta - \omega$, the error probability functions of the test ϕ may be written as follows:

$\alpha(\theta) = E_\theta(\phi)$, for $\theta \in \omega$.
$\beta(\theta) = 1 - E_\theta(\phi)$, for $\theta \in \Theta - \omega$.

We may still apply the Neyman–Pearson criteria of optimality to this more general class of tests, and, among these satisfying

$\alpha(\theta) \leqslant \alpha$,

look for that (if it exists) which uniformly maximizes $1 - \beta(\theta)$. In particular if ω is simple, $\omega = \{\theta_0\}$, and so is $\Theta - \omega$, $\Theta - \omega = \{\theta_1\}$, we may still define an optimum test as a most-powerful test of significance level α. If we do so, it is possible to prove a more general version of the fundamental lemma which establishes the existence, *in all circumstances*, of such an optimum test of a simple hypothesis against a simple alternative. This optimum test is a *randomized likelihood ratio test* which has the form

$$\phi(x) = 1, \text{ if } \frac{p_1(x)}{p_0(x)} > k,$$

$$\phi(x) = c, \text{ if } \frac{p_1(x)}{p_0(x)} = k,$$

$$\phi(x) = 0, \text{ if } \frac{p_1(x)}{p_0(x)} < k.$$

Thus in the example above, where x is the number of successes in three independent trials, and we wish to test that the probability of success is $\theta_0 = \frac{1}{4}$ against the alternative that it is $\theta_1 = \frac{3}{4}$, this optimum test would read:

if $x = 3$, decide that the probability of success θ is $\frac{3}{4}$;

if $x = 2$, decide, with probability 2·2/9, that $\theta = \frac{3}{4}$ and, with probability 6·8/9, that $\theta = \frac{1}{4}$;

if $x < 2$, decide that $\theta = \frac{1}{4}$.

The probability 2·2/9 involved here requires some explanation, which illustrates the general method of constructing optimum randomized tests. We start by considering that x for which $p_1(x)/p_0(x)$ is a maximum, namely $x = 3$. Since $P_{\theta_0}(x = 3) < 0·05$, we take $\phi(3) = 1$. Now consider that value of x for which $p_1(x)/p_0(x)$ takes its next-to-maximum value, namely $x = 2$. $P_{\theta_0}(x = 2) = \frac{9}{64}$ and since $P_{\theta_0}(x = 2 \text{ or } 3) > 0·05$, we cannot 'put the whole point $x = 2$' into the critical region and finish with a test of significance level α. So we put a 'fraction of the point $x = 2$' into the critical region, a fraction just sufficient to make the probability of the Type I error equal to 0·05. We choose c so that

$$P_{\theta_0}(x = 3) + c P_{\theta_0}(x = 2) = 0·05.$$

This equation has the solution $c = 2·2/9$, and so the test

$$\phi(3) = 1$$

$$\phi(2) = \frac{2·2}{9}$$

$$\phi(x) = 0, \text{ if } x < 2,$$

has Type I error probability 0·05 exactly.

For the version of the Neyman–Pearson lemma which establishes that this method of construction leads in general to a most-powerful test of significance level α of a simple null-hypothesis against a simple alternative, the reader is referred to Lehmann (1959), p. 65.

6.2.4   We conclude our discussion of tests of a simple null-hypothesis against a simple alternative by considering a continuous example. Let $x = (x_1, x_2, \ldots, x_n)$ be a random sample from a normal distribution with variance 1. Construct a most-powerful, α-level, test of the null-hypothesis that the mean of this distribution is θ_0 against the alternative that it is θ_1, a number greater than θ_0. Consider a set of the form $\{x : p_1(x)/p_0(x) \geq k\}$.

We have
$$\frac{p_1(x)}{p_0(x)} \geq k \Leftrightarrow \exp\{-\tfrac{1}{2}\sum(x_i-\theta_1)^2 + \tfrac{1}{2}\sum(x_i-\theta_0)^2\} \geq k$$
$$\Leftrightarrow n\bar{x}(\theta_1-\theta_0) - \tfrac{1}{2}n(\theta_1^2-\theta_0^2) \geq k'$$
$$\Leftrightarrow \bar{x} \geq k''.$$

Hence we can find a k such that $P_{\theta_0}\{x : p_1(x)/p_0(x) \geq k\} = \alpha$ iff we can find a k'' such that $P_{\theta_0}\{x : \bar{x} \geq k''\} = \alpha$. This we can certainly do. If θ_0 is the true parameter, $\bar{x}$ is $N(\theta_0, n^{-1})$, or $\sqrt{n}(\bar{x}-\theta_0)$ is $N(0, 1)$. Let k_α be the upper 100α per cent point of an $N(0, 1)$ distribution. Then by definition,
$$P_{\theta_0}\{\sqrt{n}(\bar{x}-\theta_0) \geq k_\alpha\} = \alpha.$$
Therefore $P_{\theta_0}\left\{\bar{x} \geq \dfrac{k_\alpha}{\sqrt{n}} + \theta_0\right\} = \alpha,$

that is, the set

$$R^* = \left\{x : \bar{x} \geq \frac{k_\alpha}{\sqrt{n}} + \theta_0\right\}$$

is a set satisfying the conditions required in the statement of the fundamental lemma, and so it is the critical region of a most-powerful α-level test of $\{\theta_0\}$ against $\{\theta_1\}$.

It should be noted that this critical region is not quite unique. We can add to it or subtract from it a set of probability zero under each hypothesis, and the resulting region will have the same probabilistic properties as R^*. Thus, for example, the region

$$\left\{x : \bar{x} > \frac{k_\alpha}{\sqrt{n}} + \theta_0\right\}$$

also is the critical region of a most powerful α-level test of $\{\theta_0\}$ against $\{\theta_1\}$, since, for $i = 0, 1$

$$P_{\theta_i}\left\{\bar{x} = \frac{k_\alpha}{\sqrt{n}} + \theta_0\right\} = 0.$$

6.3 Composite hypotheses

In practice almost invariably we are interested in tests of *composite* rather than simple hypotheses, a composite hypothesis ω being one where the set ω contains more than one element. Hence it is the existence of *uniformly*-most-powerful tests which will be of practical interest. Unfortunately it is the exception rather than the rule that a U.M.P. test exists.

Suppose that we consider the problem of testing a simple null hypothesis

$\{\theta_0\}$ against a composite alternative $\Theta - \{\theta_0\}$. (For instance, we might be interested in testing whether an unknown probability of success is a half against the alternative that it is not a half.) The power $1 - \beta(\theta)$ of a U.M.P. α-level test of $\{\theta_0\}$ against $\Theta - \{\theta_0\}$ must equal the power of a most-powerful α-level test of $\{\theta_0\}$ against $\{\theta\}$, for every θ in $\Theta - \{\theta_0\}$. As often as not the critical region of a most-powerful α-level test of $\{\theta_0\}$ against $\{\theta\}$ is essentially (that is, apart from zero-probability sets) unique. Hence in order that a U.M.P. test α-level test of $\{\theta_0\}$ against $\Theta - \{\theta_0\}$ exist, it is necessary that the most-powerful α-level test of $\{\theta_0\}$ against $\{\theta\}$ should be the same for all $\theta \in \Theta - \{\theta_0\}$.

This sometimes happens. Consider the example of section 6.2.4, with the problem of testing $\{\theta_0\}$ against $\{\theta_1\}$ replaced by that of testing $\{\theta_0\}$ against $\{\theta : \theta > \theta_0\}$. For *any* $\theta > \theta_0$ the test with critical region

$$\left\{ x : \bar{x} > \frac{k_\alpha}{\sqrt{n}} + \theta_0 \right\}$$

is a most-powerful α-level test of $\{\theta_0\}$ against $\{\theta\}$. Hence this test is a U.M.P. α-level test of $\{\theta_0\}$ against $\{\theta : \theta > \theta_0\}$.

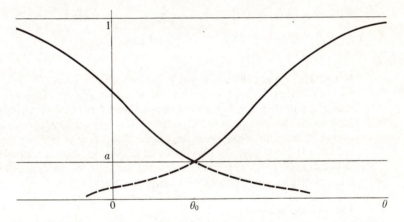

Figure 4 Graphs of power functions of size-α tests of the hypothesis that $\theta = \theta_0$

However we do not have to look much further to discover a situation where no U.M.P. test exists. Suppose that we consider the same observational situation but now wish to test $\{\theta_0\}$ against $\{\theta : \theta \neq \theta_0\}$ – what is termed a two-sided alternative, for obvious reasons. Essentially the only test which achieves maximum power at values of $\theta > \theta_0$ is that just quoted, namely that with

critical region
$$C_1 = \left\{x : \bar{x} \geqslant \frac{k_\alpha}{\sqrt{n}} + \theta_0\right\}.$$
On the other hand, if we consider a most-powerful α-level test of $\{\theta_0\}$ against $\{\theta\}$ for any $\theta < \theta_0$ we discover that such a test has essentially the critical region
$$C_2 = \left\{x : \bar{x} \leqslant -\frac{k_\alpha}{\sqrt{n}} + \theta_0\right\},$$
and the first test has much smaller power at values of $\theta < \theta_0$ than has the second. So no U.M.P. test of $\{\theta_0\}$ against $\{\theta : \theta = \theta_0\}$ exists. The position may be summarized graphically as in figure 4.

The unbroken curve in Figure 4 represents the maximum power attainable by an α-level test at different values of θ. The curve broken to the left of θ_0 and unbroken to the right represents the power function of the test with critical region C_1; while the curve dotted to the right of θ_0 and unbroken to the left represents the power function of the test C_2.

It may appear to the reader that there is an intuitively obvious best α-level test of $\{\theta_0\}$ against $\{\theta : \theta \neq \theta_0\}$, namely that with critical region
$$C_3 = \{x : |\bar{x} - \theta_0| > k\},$$
where k is determined by
$$P_{\theta_0}\{|\bar{x} - \theta_0| > k\} = \alpha.$$
Indeed this is so and this test would in fact be used in practice. The fact that the test C_3 cannot be justified as being U.M.P. is an indictment of the criterion rather than the test. We are back in a situation very familiar to the theoretical statistician. The criterion 'a U.M.P. α-level test is optimum' has considerable appeal, at least for some testing problems. However it is not helpful if a U.M.P. α-level test does not exist. More often than not this is the case, and, as we have just seen, this criterion may fail to justify an intuitively acceptable test. This suggests modification of the criterion.

6.4 Unbiased and invariant tests

6.4.1 *Unbiased tests*

When considering optimum estimators in chapter 2, we found that the criterion of 'minimum mean-square error' was not a very useful one because seldom in practice does a uniformly minimum mean-square error estimator exist. Accordingly we imposed a restriction on the class of estimators considered, a restriction designed to eliminate 'ridiculous' estimators and then asked whether there existed within this restricted class a minimum mean-square error estimator. We may adopt the same approach for tests, and there are two main ways in which this has been done.

Already we have restricted attention to α-level tests, that is, tests of ω against $\Theta - \omega$ with the property that

$$\alpha(\theta) \leqslant \alpha \quad \text{for all } \theta \in \omega,$$

where $\alpha(\theta)$ is the Type I error probability function; this on the grounds that the Type I error is more serious than the Type II. After a little thought it can be seen that it is not unnatural to impose the further restriction that the power $1 - \beta(\theta)$ of any test worth consideration should be greater than or equal to α, for all $\theta \in \Theta - \omega$. If we do not make this additional demand then we face the possibility of using a test with the property that according to it we are more liable to accept $\Theta - \omega$ when it is false than when it is true.

A test satisfying

$$\alpha(\theta) \leqslant \alpha \quad \text{for all } \theta \in \omega,$$
$$1 - \beta(\theta) \geqslant \alpha \quad \text{for all } \theta \in \Theta - \omega$$

is said to be an *unbiased* α-level test. It may be that there exists a U.M.P. unbiased α-level test, where there does not exist a U.M.P. α-level test. Indeed many intuitively acceptable tests such as the test C_3 of the preceding section can be justified as being U.M.P. unbiased α-level. However we shall not pursue the investigation of the class of unbiased tests. An excellent account of the theory and application of unbiasedness of tests is given by Lehmann (1959).

6.4.2 *Invariant tests*

For certain problems another natural restriction on the class of tests worth considering suggests itself. We shall illustrate this by means of an example rather than by a general formal mathematical statement, which requires considerable preliminary background.

Suppose that $x = (x_1, x_2)$; that x_1 and x_2 are independent and have $N(\theta_1, 1)$ and $N(\theta_2, 1)$ distributions; and that we wish to test the hypothesis that $\theta_1 = \theta_2$ against the alternative that $\theta_1 \neq \theta_2$. It seems natural to demand that the critical region of any test worthy of consideration should be symmetric in x_1 and x_2. If we do make this demand then, for instance, any test with critical region of the form $\{x; x_1 - x_2 > k\}$ would be eliminated, as it does not have the required symmetry. On the other hand any test with critical region of the form $\{x; |x_1 - x_2| > k\}$ does have the required symmetry. It may be that within the class of 'symmetric' tests there is one which is U.M.P. α-level.

Generally speaking symmetry is expressed in terms of invariance under some group of transformations, and we are thus led to consideration of the existence of U.M.P. invariant α-level tests. Again several important intuitively acceptable tests which are not U.M.P. can be justified as being U.M.P. invariant, and again we refer the reader to Lehmann (1959) for a comprehensive account of the theory and application of invariance ideas to hypothesis-testing problems.

6.4.3 Just as in the case of estimation theory we are now faced with a dilemma. The notions of size and power of a test are extremely important for comparing the virtues of different tests. However the criterion that a U.M.P. α-level test is best, even when supplemented by further restrictions such as unbiasedness and invariance, leads to methods for constructing optimum tests only in limited circumstances. What do we do when faced by a practical problem where these methods fail? We may go on seeking new criteria which yield optimum tests for a wider class of problems in the hope that they will do so for the problem at hand. Or we may look for a general method of constructing tests which has intuitive appeal and apply this to the given problem in the hope that, while it may not yield a solution which can be justified in terms of criteria already introduced, this solution will make adequate, if not necessarily optimum, use of the information provided by our observation. In the next three chapters we discuss tests constructed from the latter point of view.

Examples

6.1 In order to test the hypothesis that an unknown probability of success is less than a half, twenty independent trials are carried out, and the hypothesis is accepted if and only if the number of successes observed is less than twelve. Draw graphs of the error probability functions of this procedure.

6.2 Let $x_1, x_2, \ldots, x_n$ be a random sample from a $N(\mu, 1)$ distribution. Suppose that the critical region $\{(x_1, x_2, \ldots, x_n) : \sqrt{n}|\bar{x}| > 2\}$ is used to test the hypothesis that $\mu = 0$. What is the size of this test? Sketch the graph of its power function.

6.3 Given a random sample $x_1, x_2, \ldots, x_n$ from a $N(\mu, 1)$ distribution consider the two size-α tests with critical regions

$$R_1 = \{(x_1, x_2, \ldots, x_n); |\bar{x}| > k_1\}$$
$$R_2 = \{(x_1, x_2, \ldots, x_n); \sum x_i^2 > k_2\}$$

of the hypothesis that $\mu = 0$. (k_1 and k_2 are such that the tests have size α.) Is one of these tests uniformly more powerful than the other?

6.4 Let $x_1, x_2, \ldots, x_n$ be a random sample from the distribution on the positive real numbers with density $[\theta^q/\Gamma(q)] x^{q-1} e^{-\theta x}$. If q is known construct the most-powerful size-α test of the hypothesis $\{\theta_0\}$ against the alternative $\{\theta_1\}$, where $\theta_1 > \theta_0$, and show that there exists a U.M.P. test of $\{\theta_0\}$ against $\{\theta : \theta > \theta_0\}$. In the case where $q = 1/n$, show that the power function of this test is $1 - (1-\alpha)^{\theta/\theta_0}$.

6.5 It is claimed by the seller of fishing rights that a lake contains at least N fish, where N is a large number. To investigate this claim part of the lake is netted, and m captured fish are tagged and returned to the lake. Subsequently, when the tagged fish have distributed themselves over the lake, n fish are captured and r of these are found to be tagged. The seller's claim is rejected if r/n is greater than some number k. Show how to choose k so that the probability of falsely rejecting the claim is not more than 0·1. (You may assume that m and n are small relative to N.)

6.6 Let $x_1, x_2, \ldots, x_n$ be a random sample from the uniform distribution on $(0, \theta)$. Show that there exists a U.M.P. size-α test of $\{\theta_1\}$ against $\{\theta : \theta < \theta_1\}$. Is there a U.M.P. size-α test of $\{\theta_1\}$ against $\{\theta : \theta \neq \theta_1\}$? (*Camb. Dip.*)

6.7 The common distribution of independent, identically distributed random variables $x_1, x_2, \ldots, x_n$ has density

$$\exp\{-(x-\theta)\} \quad (x > \theta).$$

To test the null hypothesis $\{\theta : \theta \leq 1\}$ against the alternative $\{\theta : \theta > 1\}$, a critical region of the form

$$\{(x_1, x_2, \ldots, x_n) : \min(x_1, x_2, \ldots, x_n) > c\}$$

is proposed. Determine c so that this test has size α. Sketch the graph of the power function of the test.

6.8 The probability that r particles are observed in the course of a certain experiment is $e^{-\lambda}\lambda^r/r!$ ($r = 0, 1, 2, \ldots$). Prove that the probability that altogether N particles are observed in n independent replicates of the experiment is $e^{-n\lambda}(n\lambda)^N/N!$.

Suppose that λ is known to be either $\tfrac{1}{2}$ or 1. Compare the following rules for deciding, on the results of five independent replicates of the experiment which value λ has.

Rule 1. Decide that $\lambda = \tfrac{1}{2}$ if and only if the total number of particles observed is less than four.
Rule 2. Decide that $\lambda = \tfrac{1}{2}$ if and only if, in more than two replicates no particles are observed.

6.9 The random variables $x_1, x_2, \ldots, x_n$ are independent and x_i is $N(\theta_i, 1)$. Show that the most-powerful size-0·05 test of the null hypothesis that each θ_i is zero against the alternative that $\theta_i = \tfrac{1}{2}$ for $i = 1, 2, \ldots, r$ and $\theta_i = -\tfrac{1}{2}$ for $i = r+1, \ldots, n$ has critical region

$$\left\{(x_1, x_2, \ldots, x_n) : \sum_{i=1}^{r} x_i - \sum_{i=r+1}^{n} x_i > 1 \cdot 645 \sqrt{n}\right\}.$$

How large must n be to ensure that the power of this test is at least 0·9?

7 The Likelihood-Ratio Test and Alternative 'Large-Sample' Equivalents of It

Hypothesis testing problems are divided into two main classes called respectively parametric and non-parametric problems. Parametric problems are those for which the true distribution on the sample space may be assumed to be known apart from the values of a finite number of unknown real parameters – in other words when the parameter space Θ may be taken as a subset of a finite dimensional Euclidean space. In non-parametric problems the family of possible distributions is larger. For instance an observation may consist of a random sample from a distribution about which we are prepared to assume nothing other than its absolute continuity; there is then a probability distribution on the sample space corresponding to each probability density function on the line, and this family of distributions is too large to be parametrized by a finite-dimensional vector-valued parameter.

The first general method of test construction which we consider is applicable mainly to parametric problems.

7.1 The likelihood-ratio test

Suppose that the family $\{P_\theta : \theta \in \Theta\}$ of possible distributions on the sample space is defined by a family $\{p(\cdot, \theta) : \theta \in \Theta\}$ of density functions with respect to some fixed measure. Again this will usually be either 'counting' measure, in which case $p(x, \theta)$ is simply the probability of the observation x when θ is the true parameter; or Lebesgue measure when the sample space is Euclidean, and in this case $p(x, \theta)$ is the probability density at x when θ is the true parameter. We wish to test a null-hypothesis ω against the alternative $\Theta - \omega$, a terminology which contains implications regarding the seriousness of errors and suggests limiting consideration to tests of some given significance-level α.

The informal argument underlying the (generalized) likelihood-ratio test is as follows. For an observation x, determine its best chance under $\Theta - \omega$, and its best chance under ω. If the ratio of these best chances is big enough, this indicates that $\Theta - \omega$ is true and x is put into the critical region; if the ratio is not very big the observation x does not provide enough evidence to refute ω and x is not then put into the critical region. The critical value of the ratio is determined by considerations of the size of the test.

Formally the test is defined as follows:

Let $\lambda(x) = \dfrac{\sup\limits_{\theta \in \Theta - \omega} p(x, \theta)}{\sup\limits_{\theta \in \omega} p(x, \theta)}.$

The critical region of the size-α likelihood-ratio test of ω against $\Theta - \omega$ is

$\{x : \lambda(x) > k\}$

where k is determined by the condition

$\sup\limits_{\theta \in \omega} P_\theta\{x : \lambda(x) > k\} = \alpha.$

(In the case of discrete probability distributions – see section 6.2 – it may not be possible to find a non-randomized test whose size is exactly equal to α; in this case k is chosen to make the size of the test as nearly as possible α, consistent with its significance level being α.)

7.1.1 Example

An observation x is the number of successes in n independent trials with unknown probability θ of success in each. Find an α-level likelihood-ratio test of the hypothesis that $\theta \leqslant \theta_0$ against the alternative that $\theta > \theta_0$.

If $\dfrac{x}{n} \leqslant \theta_0$, then $\lambda(x) = \dfrac{p(x, \theta_0)}{p(x, x/n)} = \dfrac{\theta_0^x (1-\theta_0)^{n-x}}{(x/n)^x (1-x/n)^{n-x}} \leqslant 1.$

If $\dfrac{x}{n} > \theta_0$, then $\lambda(x) = \dfrac{p(x, x/n)}{p(x, \theta_0)} = \dfrac{(x/n)^x (1-x/n)^{n-x}}{\theta_0^x (1-\theta_0)^{n-x}}.$

Now $\lambda(x)$ is an increasing function of x and so $\lambda(x) > k$ iff $x > k'$. Hence the critical region of a likelihood-ratio test takes the form $\{x : x > k'\}$.

We have to choose k' so that, if possible,

$\sup\limits_{\theta \leqslant \theta_0} P_\theta\{x > k'\} = \alpha.$

It is clear that

$\sup\limits_{\theta \leqslant \theta_0} P_\theta\{x > k'\} = P_{\theta_0}\{x > k'\}.$

Hence we must choose k' so that, if possible,

$P_{\theta_0}\{x > k'\} = \alpha.$

Because of the discreteness of the problem equality may not be possible for a non-randomized test. So we choose k' to be that integer such that

$\qquad P_{\theta_0}\{x > k'\} \leqslant \alpha$

and $\quad P_{\theta_0}\{x > k'-1\} > \alpha.$

This test is of course the only intuitively reasonable α-level test of $\{\theta : \theta \leq \theta_0\}$ against $\{\theta : \theta > \theta_0\}$. According to it a large enough number of successes indicates that the alternative is true. Indeed if the size of the test happens to be α, it is a U.M.P. α-level test (see Lehmann, 1959, p. 70).

7.1.2 Example

A problem of fairly common occurrence is the following. An observation x is of the form $x = (y, z)$ where $y = (y_1, y_2, \ldots, y_n)$ and $z = (z_1, z_2, \ldots, z_n)$ are random samples from distributions which are assumed $N(\mu_1, \sigma^2)$ and $N(\mu_2, \sigma^2)$ respectively, μ_1, μ_2 and σ^2 being unknown. We wish to test the hypothesis that $\mu_1 = \mu_2$ against the alternative that $\mu_1 \neq \mu_2$. This problem arises, for instance, when the ys are measurements on control (untreated) patients and the zs corresponding measurements on treated patients; we are prepared to assume that the only possible effect of the treatment is to 'shift' the distribution mean; and we wish to decide whether we have enough evidence to conclude that it does so.

We may arrive at a 'reasonable' test by the following heuristic argument. If $\mu_1 = \mu_2$ then $\bar{y} = n^{-1} \sum y_i$ and $\bar{z} = n^{-1} \sum z_i$ will tend to be relatively close to one another; if $\mu_1 \neq \mu_2$, they will tend to be relatively further apart. Relative to what? Relative to the inherent variability in experimental units, or, in other words, to σ. So $|\bar{y} - \bar{z}|/\sigma$ will tend to be small if $\mu_1 = \mu_2$ and to be larger if $\mu_1 \neq \mu_2$. We do not know σ, but the same argument will apply if we replace σ by an estimate of it. Now an unbiased estimate of σ^2, which apparently is as good as possible is

$$s^2 = \frac{1}{2n-2}\left[\sum (y_i - \bar{y})^2 + \sum (z_i - \bar{z})^2\right],$$

the average of 'best' estimates obtained from the separate samples. Hence a reasonable-looking critical region for the test of interest might take the form

$$\left\{x : \frac{|\bar{y} - \bar{z}|}{s} > k\right\}.$$

Let us now determine the shape of the critical region of a likelihood-ratio test. In the example

$$\theta = (\mu_1, \mu_2, \sigma^2), \quad \Theta = \{\theta : -\infty < \mu_1 < \infty, -\infty < \mu_2 < \infty, \sigma^2 > 0\},$$
and $\omega = \{\theta : \theta \in \Theta, \mu_1 = \mu_2\}$.

Moreover $p(x, \theta) = \dfrac{1}{(2\pi)^n \sigma^{2n}} \exp\left[-\dfrac{1}{2\sigma^2}\left\{\sum (y_i - \mu_1)^2 + \sum (z_i - \mu_2)^2\right\}\right]$,

and $\sup_{\theta \in \Theta - \omega} p(x, \theta) = \dfrac{1}{(2\pi)^n \hat{\sigma}^{2n}} \exp\left[-\dfrac{1}{2\hat{\sigma}^2}\left\{\sum (y_i - \bar{y})^2 + \sum (z_i - \bar{z})^2\right\}\right]$

$= \dfrac{1}{(2\pi)^n \hat{\sigma}^{2n}} e^{-n},$

where $\hat{\sigma}^2 = \dfrac{1}{2n}\left[\sum(y_i-\bar{y})^2 + \sum(z_i-\bar{z})^2\right]$;

while $\sup\limits_{\theta\in\omega} p(x,\theta) = \dfrac{1}{(2\pi)^n \dot{\sigma}^{2n}} e^{-n}$,

where $\dot{\sigma}^2 = \dfrac{1}{2n}\left[\sum(y_i-\hat{\mu})^2 + \sum(z_i-\hat{\mu})^2\right]$, $\quad \hat{\mu} = \dfrac{1}{2}(\bar{y}+\bar{z})$.

Hence $\lambda(x) = \left[\dfrac{\hat{\sigma}^2}{\dot{\sigma}^2}\right]^n$.

Now $\dot{\sigma}^2 = \hat{\sigma}^2 + \tfrac{1}{4}(\bar{y}-\bar{z})^2$,

and so $\lambda(x) > k$ iff $\dfrac{|\bar{y}-\bar{z}|}{\hat{\sigma}} > k'$

and this inequality in turn holds

iff $\dfrac{|\bar{y}-\bar{z}|}{s} >$ some constant.

Hence the 'shape' of the critical region of a likelihood-ratio test is exactly that to which we were led by the previous heuristic argument.

If we wish a size-α test, it remains to determine a constant c such that

$$\sup_{\theta\in\omega} P_\theta\left\{\dfrac{|\bar{y}-\bar{z}|}{s} > c\right\} = \alpha.$$

It transpires that the θ-distribution of $(\bar{y}-\bar{z})/s$ is the same for all $\theta\in\omega$, that, in fact for all such θ, $\{(\bar{y}-\bar{z})/s\}\sqrt{(\tfrac{1}{2}n)}$ has Student's t-distribution with $2n-2$ degrees of freedom. So c can readily be determined from tables of this distribution.

This is a case where the likelihood-ratio test can be justified in terms of criteria introduced in chapter 6. It is U.M.P. unbiased and also U.M.P. invariant (Lehmann, 1959, pp. 172, 224).

Note. The likelihood-ratio size-α test discussed in this section has the property that

$\alpha(\theta) = \alpha, \quad$ for all $\theta\in\omega$,

where $\alpha(\theta)$ is the Type I error probability function. Generally a test with this property is said to be *similar*.

7.1.3 Example: testing the hypothesis of no association in a contingency table

We have already discussed (chapter 6) the formal translation of this hypothesis into the general terms which we are adopting throughout. An observation x has the form $(n_{11}, n_{12}, \ldots, n_{ab})$ where the n_{ij}s are integers. A typical θ has the form $(\theta_{11}, \theta_{12}, \ldots, \theta_{ab})$ where $0 \leq \theta_{ij} \leq 1$ and $\sum_{i,j} \theta_{ij} = 1$,

and $p(x, \theta) = \text{constant} \times \prod_{i,j} \theta_{ij}^{n_{ij}}$.

The null hypothesis ω is the hypothesis $\{\theta : \theta \in \Theta, \theta_{ij} = \theta_{i\cdot}\theta_{\cdot j}, \text{ all } i \text{ and } j\}$ where $\theta_{i\cdot} = \sum_j \theta_{ij}$ and $\theta_{\cdot j} = \sum_i \theta_{ij}$.

For any $\theta \in \omega$ we have

$$p(x, \theta) = \text{constant} \times \prod_{i,j} (\theta_{i\cdot}\theta_{\cdot j})^{n_{ij}}$$

$$= \text{constant} \times \prod_i \theta_{i\cdot}^{n_{i\cdot}} \prod_j \theta_{\cdot j}^{n_{\cdot j}} \quad \text{in obvious notation,}$$

and remembering that $\sum_i \theta_{i\cdot} = \sum_j \theta_{\cdot j} = 1$, we find that

$$\sup_{\theta \in \omega} p(x, \theta) = \text{constant} \times \prod_i \left[\frac{n_{i\cdot}}{n_{\cdot\cdot}}\right]^{n_{i\cdot}} \prod_j \left[\frac{n_{\cdot j}}{n_{\cdot\cdot}}\right]^{n_{\cdot j}}.$$

Obviously $\sup_{\theta \in \Theta - \omega} p(x, \theta) = \text{constant} \times \prod_{i,j} \left[\frac{n_{ij}}{n_{\cdot\cdot}}\right]^{n_{ij}}$.

Hence $\lambda(x) = n_{\cdot\cdot}^{n_{\cdot\cdot}} \prod_{i,j} \left[\frac{n_{ij}}{n_{i\cdot} n_{\cdot j}}\right]^{n_{ij}}$,

so that $\log \lambda(x) = \sum_{i,j} n_{ij} \log n_{ij} - \sum_i n_{i\cdot} \log n_{i\cdot} - \sum_j n_{\cdot j} \log n_{\cdot j} + n_{\cdot\cdot} \log n_{\cdot\cdot}$.

It is apparent that there is little difficulty in determining the *shape* of the critical region of a likelihood-ratio test in this case. However, suppose that we wish a size-α test. Then we have the problem of choosing k_α so that

$$\sup_{\theta \in \omega} P_\theta \{\lambda(x) > k_\alpha\} = \alpha.$$

Now $\lambda(x)$ is rather a complicated function of the observed random variables – the n_{ij}s – and the problem of distribution calculus involved in determining its θ-distribution is far from trivial. But if $n_{\cdot\cdot}$ is large we can appeal to the following important large sample result which enables us to obtain a good approximation to the appropriate value of k_α.

7.2 The large-sample distribution of λ

7.2.1

Suppose that an observation x is a random sample from some distribution, that is, $x = (x_1, x_2, \ldots, x_n)$ where the x_is are independent and identically distributed. Suppose further that the true distribution on the space of a single observation is known apart from the values of a finite number s of unknown real parameters; that is, the parameter space Θ is a subset of R^s. We wish to test a hypothesis ω which is specified by imposing r restrictions on

$$\theta = (\theta_1, \theta_2, \ldots, \theta_s),$$

say the restrictions $h_j(\theta) = 0, j = 1, 2, \ldots, r$. Here we assume that there are no redundant restrictions, and that θ is identifiable without any restrictions, so that none of the stated restrictions are required for identifiability. In addition suppose that the family $\{P_\theta^* : \theta \in \Theta\}$ of distributions on the space of a single observation is defined by a family $\{p_\theta^* : \theta \in \Theta\}$ of density functions with respect to some fixed measure. This means that there is a family $\{p(\cdot, \theta) : \theta \in \Theta\}$ of density functions on our sample space of n observations and

$$p(x, \theta) = \prod_{i=1}^{n} p_\theta^*(x_i).$$

Typically with this set-up we shall have

$$\sup_{\theta \in \Theta - \omega} p(x, \theta) = p\{x, \hat{\theta}(x)\},$$

where $\hat{\theta}$ is an unrestricted maximum-likelihood estimator of θ,

while $\sup_{\theta \in \omega} p(x, \theta) = p\{x, \check{\theta}(x)\},$

where $\check{\theta}$ is a restricted maximum-likelihood estimator of θ, restricted by the conditions $h_j(\theta) = 0, j = 1, 2, \ldots, r$; and in this case

$$\lambda(x) = \frac{p\{x, \hat{\theta}(x)\}}{p\{x, \check{\theta}(x)\}}.$$

The important large-sample result of distribution calculus to which we have referred is:

7.2.2 Theorem

In the situation just described, subject to regularity, $2 \log \lambda$ is distributed, for all $\theta \in \omega$, approximately as $\chi^2(r)$, r being the number of restrictions on θ required to define ω.

Proof. We shall not give a detailed proof of this result, but shall merely indicate how it follows from the large-sample distribution theory of maximum-likelihood estimators, restricted and unrestricted, which we considered in chapter 4.

Let θ_0 denote the true parameter, and suppose that $\theta_0 \in \omega$. In this case if n is large, both $\hat{\theta}$ and $\check{\theta}$ are near θ_0, and so near one another.

Consequently $2 \log \lambda = 2\{\log p(x, \hat{\theta}) - \log p(x, \dot{\theta})\}$
$$= (\hat{\theta} - \dot{\theta})'\{-D_\theta^2 \log p(x, \hat{\theta})\}(\hat{\theta} - \dot{\theta}),$$

a result obtained by expanding $\log p(x, \dot{\theta})$ in a Taylor's series about $\hat{\theta}$, and assuming that $D_\theta \log p(x, \hat{\theta}) = 0$, that is, that the unrestricted estimator $\hat{\theta}$ emerges as a solution of the likelihood equations.

Moreover $-D_\theta^2 \log p(x, \hat{\theta}) \simeq n\mathbf{B}_{\hat{\theta}} \simeq n\mathbf{B}_{\theta_0}$,

where $\mathbf{B}_\theta$ is the information matrix for a single observation.

Hence $2 \log \lambda \simeq n(\hat{\theta} - \dot{\theta})' \mathbf{B}_{\theta_0}(\hat{\theta} - \dot{\theta})$.

Now if $Y = n^{-\frac{1}{2}} D_\theta \log p(x, \theta_0)$ then Y is approximately $N(0, \mathbf{B}_{\theta_0})$ (see section 4.6)

and $\sqrt{n}(\hat{\theta} - \theta_0) \simeq \mathbf{B}_{\theta_0}^{-1} Y$,

while $\sqrt{n}(\dot{\theta} - \theta_0) \simeq \mathbf{P}_{\theta_0} Y$,

where $\mathbf{P}_{\theta_0}$ is defined in section 4.7.3.

Therefore $2 \log \lambda \simeq Y'(\mathbf{B}_{\theta_0}^{-1} - \mathbf{P}_{\theta_0}) Y$.

Let $\mathbf{B}_{\theta_0} = AA'$, where A is non-singular.; this being possible if the information matrix $\mathbf{B}_\theta$ is non-singular as it normally is when θ is identifiable. Further let $Y = AZ$ so that $Z = A^{-1}Y$ is approximately $N(0, I_s)$.

Then $2 \log \lambda \simeq Z'A'(\mathbf{B}_{\theta_0}^{-1} - \mathbf{P}_{\theta_0})AZ$
$$= Z'Z - Z'(A'\mathbf{P}_{\theta_0}A)Z.$$

Now $(A'\mathbf{P}_{\theta_0}A)^2 = A'\mathbf{P}_{\theta_0}AA'\mathbf{P}_{\theta_0}A = A'(\mathbf{P}_{\theta_0}\mathbf{B}_{\theta_0}\mathbf{P}_{\theta_0})A$,

and $\mathbf{P}_{\theta_0}\mathbf{B}_{\theta_0}\mathbf{P}_{\theta_0} = \mathbf{P}_{\theta_0}$ (see section 3.10.1).

Hence $A'\mathbf{P}_{\theta_0}A$ is idempotent, and its rank is that of $\mathbf{P}_{\theta_0}$, namely $s - r$. It follows that $(I_s - A'\mathbf{P}_{\theta_0}A)$ is idempotent of rank r.

Now $2 \log \lambda \simeq Z'(I_s - A'\mathbf{P}_{\theta_0}A)Z$.

The right hand side being a quadratic form in independent $N(0, 1)$ random variables with an idempotent matrix of rank r, is distributed as $\chi^2(r)$. Hence $2 \log \lambda$ is distributed approximately as $\chi^2(r)$.

7.2.3   Let us apply this result to the problem of section 7.1.3, that of testing the no-association hypothesis in a contingency table. Here the parameter space

$$\Theta = \left\{\theta = (\theta_{11}, \theta_{12}, \ldots, \theta_{ab}) : 0 \leqslant \theta_{ij} \leqslant 1, \sum_{i,j} \theta_{ij} = 1\right\},$$

has dimension $ab - 1$, for it contains an $(ab - 1)$-dimensional rectangle but not one of dimension ab. The hypothesis to be tested specifies the subset

$$\omega = \{\theta \in \Theta : \theta_{ij} = \theta_{i.}\theta_{.j}, \text{ all } i \text{ and } j\}.$$

In this definition of ω there are $a+b-2$ 'free' parameters, namely $\theta_{1\cdot}, \theta_{2\cdot},\ldots,$ $\theta_{a-1,\cdot}$, (but not $\theta_{a\cdot}$, because $\sum_i \theta_{i\cdot} = 1$) and $\theta_{\cdot 1}, \theta_{\cdot 2},\ldots,\theta_{\cdot,b-1}$ (but not $\theta_{\cdot b}$). So ω has dimension $a+b-2$. Hence we must have imposed $(ab-1)-(a+b-2)$ non-redundant restrictions on $\theta \in \Theta$ to ensure that $\theta \in \omega$. Of course
$$(ab-1)-(a+b-2) = (a-1)(b-1),$$
and so in this case if $n_{\cdot\cdot}$, the total number of individuals observed, is large, $2\log\lambda$ is distributed approximately as $\chi^2\{(a-1)(b-1)\}$, when the no-association hypothesis is in fact true. If k_α is the upper 100α per cent point of such a χ^2-distribution, then $\{x: 2\log\lambda(x) > k_\alpha\}$ is the critical region of a test of the no-association hypothesis, which is approximately similar size α.

The reader may find somewhat facile the argument by which we arrived at the conclusion that ω is obtained from Θ by imposing $(a-1)(b-1)$ 'independent' restrictions on θ. If so he may be more convinced by 'reparametrizing' the whole problem in a way which corresponds precisely with the description in section 7.2.1, and writing down explicitly the appropriate independent restrictions in terms of the new parameters.

7.2.4 In many practical hypothesis-testing problems, the parameter space Θ is a subset of $\mathbf{R}^s$ and the set ω specified by the null-hypothesis is defined by stating essentially that $\theta \in \Theta$ satisfies certain restrictions, say $h_j(\theta) = 0$, $j = 1, 2,\ldots, r$.

This basic problem may be disguised to some extent by a description, which for reasons of symmetry, necessitates certain restrictions on θ for identifiability and also by stating the restrictions, which define ω, in terms of freedom equations rather than of constraint equations. Even when disguised in such ways, however, the problems can, at least in theory, be reduced to the form stated initially by a reparametrization, and it suits our convenience to think of such problems as expressed in this 'canonical' form. Application of the likelihood-ratio test to such problems usually involves the calculation of an unrestricted maximum-likelihood estimate $\hat{\theta}(x)$, and also a restricted maximum-likelihood estimate $\dot{\theta}(x)$. In examples 7.1.2 and 7.1.3 it was easily possible to obtain closed expressions for both of these estimates. However in other problems it is often necessary to determine at least one of them by numerical solution of equations and this can be time consuming. We therefore now consider two alternatives to the likelihood-ratio test each of which involves only one of the estimates $\hat{\theta}(x)$ and $\dot{\theta}(x)$.

7.3 The W-test

An obvious way of testing a null-hypothesis ω that an unknown vector-valued parameter θ satisfies the restrictions $h_1(\theta) = h_2(\theta) = \ldots = h_r(\theta) = 0$, is to calculate $\hat{\theta}$, the unrestricted maximum-likelihood estimate of θ, and base our decision on the proximity to the zero vector of the vector
$$h(\hat{\theta}) = \{h_1(\hat{\theta}), h_2(\hat{\theta}),\ldots, h_r(\hat{\theta})\}'.$$

This idea was first exploited by Wald (1943). If $h(\hat{\theta})$ is 'near enough' to the zero vector we accept ω and if not we reject it. However what 'shape' of neighbourhood of the zero vector do we take to define an acceptance region of ω? Do we take a spherical neighbourhood, a 'rectangular' neighbourhood, or what? In general the answer to this question is not immediately obvious, but if we are dealing with large samples a natural choice emerges.

Suppose then that an observation x is a large sample $(x_1, x_2, \ldots, x_n)$ from some distribution. Then, as we know, subject to mild regularity, $\hat{\theta}$ is, with θ-probability close to 1, near θ; moreover $\sqrt{n}(\hat{\theta}-\theta)$ is approximately $N(0, B_\theta^{-1})$. On expanding the components of $h(\hat{\theta})$ about the true parameter θ, by Taylor's theorem, we have

$$h(\hat{\theta}) \simeq h(\theta) + H'_\theta(\hat{\theta}-\theta),$$

where H_θ is the $s \times r$ matrix $(\partial h_j(\theta)/\partial \theta_i)$, and if $h(\theta) = 0$, that is, if the true parameter satisfies the restrictions, then

$$h(\hat{\theta}) \simeq H'_\theta(\hat{\theta}-\theta),$$

so that $\sqrt{n}h(\hat{\theta})$ is approximately $N(0, H'_\theta B_\theta^{-1} H_\theta)$. It now becomes natural to choose a region within a constant probability contour as a neighbourhood of the origin for defining an acceptance region for the null-hypothesis; in other words to take as rejection region a set of the form

$$\{x : n[h(\hat{\theta}(x))]'[H'_\theta B_\theta^{-1} H_\theta]^{-1}[h(\hat{\theta}(x))] > k\}.$$

To the order of approximation to which we are working, θ may be replaced by $\hat{\theta}(x)$ in the above expression, so that, writing $\hat{\theta}$ instead of $\hat{\theta}(x)$ for typographical brevity, a natural choice of critical region is

$$\{x : n[h(\hat{\theta})]'[H'_{\hat{\theta}} B_{\hat{\theta}}^{-1} H_{\hat{\theta}}]^{-1}[h(\hat{\theta})] > k\},$$

and this is the critical region of a W-test. It will be seen that if $h(\theta) = 0$, then the test statistic $n[h(\hat{\theta})]'[H'_{\hat{\theta}} B_{\hat{\theta}}^{-1} H_{\hat{\theta}}]^{-1}[h(\hat{\theta})]$ is distributed approximately as $\chi^2(r)$. Therefore the problem of choosing k so that the test has approximate size α presents no difficulty, when n is large.

Note that this test is primarily a large-sample test. It may be applied when we are not dealing with large samples, but the choice of k which yields an approximate size-α test may then be difficult since we cannot appeal to large-sample theory. Indeed if we are not dealing with a large sample, there is no obvious reason why a critical region of the shape of that of the W-test should be chosen in preference to some other.

7.3.1 Example

Let $y_{i1}, y_{i2}, \ldots, y_{in}$, $i = 1, 2, 3, 4$ be independent random samples from exponential distributions with unknown scale parameters θ_1, θ_2, θ_3, θ_4 respectively. Assuming that n is large, establish the W-test of the hypothesis that $\theta_1, \theta_2, \theta_3, \theta_4$ are in geometric progression.

If we write $x_j = (y_{1j}, y_{2j}, y_{3j}, y_{4j})$

then x_j may be regarded as a single observation on a vector-valued random variable whose distribution depends on the vector-valued parameter $\theta = (\theta_1, \theta_2, \theta_3, \theta_4)$, and $x_1, x_2, \ldots, x_n$ is a random sample from this distribution on $\mathbf{R}^4$. The density function $p^*(\cdot, \theta)$ of a 'single observation' is given by

$$p^*(x_j, \theta) = \prod_{i=1}^{4} \{\theta_i \exp(-\theta_i y_{ij})\},$$

so that $\log p^*(x_j, \theta) = \sum_{i=1}^{4} \log \theta_i - \sum_{i=1}^{4} \theta_i y_{ij}.$

It is an easy matter to deduce that the information matrix,

$$\mathbf{B}_\theta = \left[E_\theta \left(-\frac{\partial^2 \log p^*}{\partial \theta_i \, \partial \theta_j} \right) \right],$$

is in this case the diagonal matrix

$$\operatorname{diag}\left[\frac{1}{\theta_1^2}, \frac{1}{\theta_2^2}, \frac{1}{\theta_3^2}, \frac{1}{\theta_4^2} \right].$$

Moreover the likelihood function $p(x, \theta)$ corresponding to the random sample $x = (x_1, x_2, \ldots, x_n)$ is

$$p(x, \theta) = \prod_j \prod_i \{\theta_i \exp(-\theta_i y_{ij})\}$$

from which it is clear that unrestricted maximum-likelihood estimates of the θ_is are given by

$$\hat{\theta}_i = y_{i\cdot}^{-1},$$

where $n y_{i\cdot} = \sum_j y_{ij} \quad (i = 1, 2, 3, 4).$

The hypothesis to be tested states that the θ_is satisfy the restrictions

$$\frac{\theta_2}{\theta_1} = \frac{\theta_3}{\theta_2} = \frac{\theta_4}{\theta_3}$$

or, equivalently, $h(\theta) = 0$,

where $h(\theta) = \begin{bmatrix} \theta_2^2 - \theta_1 \theta_3 \\ \theta_3^2 - \theta_2 \theta_4 \end{bmatrix}.$

Thus in this case the matrix $\mathbf{H}'_\theta$ of our general discussion is

$$\mathbf{H}'_\theta = \begin{bmatrix} -\theta_3 & 2\theta_2 & -\theta_1 & 0 \\ 0 & -\theta_4 & 2\theta_3 & -\theta_2 \end{bmatrix}.$$

and $\mathbf{H}'_{\hat{\theta}} \mathbf{B}_{\hat{\theta}}^{-1} \mathbf{H}_{\hat{\theta}} = \begin{bmatrix} \dfrac{4}{y_2^4} - \dfrac{2}{y_1^2 \cdot y_3^2} & -\dfrac{2}{y_4 \cdot y_2^3} - \dfrac{2}{y_1 \cdot y_3^3} \\ \dfrac{1}{y_4 \cdot y_2^3} - \dfrac{2}{y_1 \cdot y_3^3} & -\dfrac{2}{y_2^2 \cdot y_4^2} + \dfrac{4}{y_3^4} \end{bmatrix}.$

Final calculation of the test statistic is now a matter of substitution in the general formula and this is left to the reader.

This example illustrates the point that while a W-test is a large-sample equivalent of a likelihood-ratio test, it may involve considerably less computation. Here calculation of restricted maximum-likelihood estimates would involve considerable work, whereas the unrestricted estimates are easy to calculate.

7.4 The χ^2 test

In other problems of the general nature under discussion, it is relatively easy to compute restricted estimates, and it is therefore desirable to have a test based solely on restricted estimates as an alternative to the likelihood-ratio test. The so-called chi-squared test provides such an alternative, for large samples.

7.4.1 The general problem is as before: we have avilable a large sample from a distribution which is known apart from the value of a vector-valued parameter θ, and we wish to test a hypothesis which states that this parameter satisfies certain restrictions, which we represent in canonical form by $h(\theta) = 0$, though in practice they may be specified in terms of freedom equations. The idea underlying the χ^2-test is this: if the hypothesis is true then a restricted maximum-likelihood estimate of θ will tend to be very near an unrestricted M.L.E., for large samples, and consequently, if we assume regularity of the log-likelihood function, the partial derivatives of this function at the restricted maximum will tend to be small. If on the other hand the hypothesis is false there is no obvious reason for these partial derivatives to be small. Hence we may use the proximity to the zero-vector of the vector

$$D_\theta \log p(x, \theta) = \left[\frac{\partial}{\partial \theta_i} \log p(x, \theta) \right],$$

as a means of deciding between the truth or falseness of the hypothesis to be tested. Again it is not quite clear what metric we should use, but for large samples a natural choice emerges.

On expanding $D_\theta \log p(x, \theta)$ about θ, we have, assuming that the true parameter satisfies the restrictions, so that $\hat{\theta}$ and θ are probably close to one another when n is large,

$$D_\theta \log p(x, \theta) \simeq D_\theta \log p(x, \hat{\theta}) + [D_\theta^2 \log p(x, \hat{\theta})](\theta - \hat{\theta}),$$

where $D_\theta^2 \log p(x, \hat\theta)$ is the matrix

$$\left[\frac{\partial^2}{\partial \theta_i\, \partial \theta_j} \log p(x, \theta)\right].$$

Now if $\hat\theta$ emerges as a solution of the likelihood equations, then

$D_\theta \log p(x, \hat\theta) = 0.$

Also, by an argument already used in large-sample maximum-likelihood theory, $D_\theta^2 \log p(x, \hat\theta) \simeq -n\mathbf{B}_{\hat\theta}$,

so that $\quad D_\theta \log p(x, \theta) \simeq n\mathbf{B}_{\hat\theta}(\hat\theta - \theta).$

Let $\quad \chi^2 = \dfrac{1}{n}[D_\theta \log p(x, \theta)]' \mathbf{B}_{\hat\theta}^{-1} [D_\theta \log p(x, \theta)].$

Then $\quad \chi^2 \simeq \dfrac{1}{n}[D_\theta \log p(x, \theta)]' \mathbf{B}_{\hat\theta}^{-1} [D_\theta \log p(x, \theta)]$

$\qquad\qquad = n(\hat\theta - \theta)' \mathbf{B}_{\hat\theta}(\hat\theta - \theta)$

$\qquad\qquad \simeq 2 \log \lambda,$

by the argument of section 7.2.2, where λ is the likelihood-ratio test statistic.

When the hypothesis under test is true, the statistic χ^2 is therefore a large-sample approximation to $2 \log \lambda$, an approximation which depends only on the restricted M.L.E. $\hat\theta$. Consequently if we are going to base a test on the proximity to zero of $D_\theta \log p(x, \theta)$, it is natural to choose as critical region of this test

$\{x : \chi^2 > k\},$

and such a test is called a χ^2 test.

The relationship between the powers of a size-α chi-squared test and a size-α likelihood-ratio test is not immediately apparent since we have not discussed the relative values of χ^2 and $2 \log \lambda$ when the hypothesis under test is not true. However subject to regularity it can be shown that the two tests have approximately the same power function for large samples (Silvey, 1959), and so for most practical purposes they are large-sample equivalents of one another. We may choose that which is most convenient from a computational point of view.

A practical difficulty sometimes arises because it happens to be convenient for reasons of symmetry to parametrize the family of possible distributions on the sample space in such a way that θ is not identifiable without restrictions. Then it is usually the case that $\mathbf{B}_\theta$ is singular. The statistic of the chi-squared test involves $\mathbf{B}_{\hat\theta}^{-1}$ so that apparently it is undefined when $\mathbf{B}_{\hat\theta}$ is singular. This is a purely technical problem which we can always overcome by reparametrization, but it is not even necessary to do this. A simple adjustment to the statistic takes care of this contingency, and at the same time preserves the symmetry which is usually the cause of the difficulty. Suppose that θ is not identifiable without restrictions and that, of the restrictions

$$h_1(\theta) = h_2(\theta) = \ldots = h_r(\theta) = 0,$$

the first t are just sufficient to make θ identifiable. Then we need only replace B_θ by $B_\theta + H_{1\theta} H'_{1\theta}$ as in section 4.7.5 in order to define the chi-squared statistic which then becomes

$$\chi^2 = \frac{1}{n} [D_\theta \log p(x, \theta)]' [B_{\hat\theta} + H_{1\hat\theta} H'_{1\hat\theta}]^{-1} [D_\theta \log p(x, \theta)],$$

and which is distributed, when the null-hypothesis is true, as $\chi^2(r-t)$.

7.4.2 When the family of possible distributions on the sample space is multinomial, the statistic of a chi-squared test reduces to a simple intuitively appealing form and possibly partly for this reason, the multinomial family of distributions provides the richest source of applications of the chi-squared test. We shall content ourselves here with a statement of the result which the interested reader may derive for himself from the general expression given above.

Suppose that each individual in a very large population belongs to exactly one of s classes and that the proportions $\theta_1, \theta_2, \ldots, \theta_s$ of individuals in these classes are unknown. We have available a large random sample of n individuals and $n_1, n_2, \ldots, n_s$ of these fall respectively into the classes $1, 2, \ldots, s$, where $n_1 + n_2 + \ldots + n_s = n$. If $x = (n_1, n_2, \ldots, n_s)$ the likelihood function is then defined by

$$p(x, \theta) = \frac{n!}{n_1! \, n_2! \ldots n_s!} \prod_{i=1}^{s} \theta_i^{n_i}, \quad \text{where} \sum_{i=1}^{s} \theta_i = 1.$$

Suppose now that we wish to test whether the θ_is satisfy the additional $(r-1)$ restrictions

$$h_1(\theta_1, \theta_2, \ldots, \theta_s) = h_2(\theta_1, \theta_2, \ldots, \theta_s) = \ldots = h_{r-1}(\theta_1, \theta_2, \ldots, \theta_s) = 0.$$

To apply a chi-squared test we calculate restricted M.L.E.s, $\hat\theta_1, \hat\theta_2, \ldots, \hat\theta_s$ of $\theta_1, \theta_2, \ldots, \theta_s$ respectively and the test statistic reduces to

$$\chi^2 = \sum_{i=1}^{s} \frac{(n_i - n\hat\theta_i)^2}{n\hat\theta_i},$$

or, as it is sometimes expressed,

$$\chi^2 = \sum \frac{(\text{observed} - \text{expected})^2}{\text{expected}},$$

the expected numbers in the various classes being calculated as if the restricted estimates of the θ_is were the true values of these parameters. When the null-hypothesis is true this statistic is distributed approximately as $\chi^2(r-1)$ and this result enables us easily to determine a test which is approximately similar size α, namely the test with critical region

$$\{x : \chi^2 > k_\alpha\},$$

where k_α is the upper 100α per cent point of a $\chi^2(r-1)$ distribution.

Examples

7.1 Let $x_1, x_2, \ldots, x_n$ be a random sample from a $N(\mu, \sigma^2)$ distribution with μ and σ^2 unknown. Show that a likelihood-ratio test of the null-hypothesis that $\sigma = \sigma_0$ against the alternative that $\sigma \neq \sigma_0$ has an acceptance region of the form

$$\left\{(x_1, x_2, \ldots, x_n) : k_1 \leqslant \frac{s^2}{\sigma_0^2} \leqslant k_2\right\},$$

where $s^2 = \sum(x_i - \bar{x})^2/n$; and explain how k_1 and k_2 are determined to make the test of size α.

7.2 Let $x_1, x_2, \ldots, x_m$ and $y_1, y_2, \ldots, y_n$ be independent random samples from two exponential distributions with unknown scale parameters λ and μ respectively. Show that the critical region of a likelihood-ratio test of the null-hypothesis that $\lambda = \mu$ against the alternative that $\lambda \neq \mu$ depends only on the ratio $\bar{y}/\bar{x}$. Explain how to make the test of size α.

7.3 An experiment has $N+1$ possible outcomes $z_0, z_1, \ldots z_N$. A null-hypothesis H_0 assigns probabilities to these as follows:

$$P(z_0) = \tfrac{1}{2} \qquad P(z_i) = \frac{1}{2N} \quad (i = 1, 2, \ldots, N).$$

A composite alternative H_1 assigns probability $(N-1)/N$ to z_0 and does not specify the probabilities of $z_1, z_2, \ldots z_N$. Show that the size-$\tfrac{1}{2}$ likelihood-ratio test of H_0 against H_1, based on a single observation, accepts H_0 if and only if the observation is z_0. What is the power of this test? Is it a 'good' test?

7.4 Let $x = (x_1, x_2, \ldots, x_n)$ be a random sample from an $N(\mu, \sigma^2)$ distribution, where μ and σ^2 are unknown, and let H_0 be the hypothesis that $\mu = \mu_0$. If $\Lambda(x)$ is the likelihood-ratio statistic for testing H_0 against the alternative that $\mu \neq \mu_0$, show that

$$2 \log \Lambda(x) = n \log\left[1 + \frac{n(\bar{x}-\mu_0)^2}{\sum(x_i - \bar{x})^2}\right].$$

Determine the characteristic function of $2 \log \Lambda(x)$ when H_0 is true, and verify that as $n \to \infty$, the distribution of $2 \log \Lambda(x)$ tends to χ^2 with one degree of freedom. Check the steps of Theorem 7.2.2 in terms of this example.

7.5 Sets of independent trials are conducted on k different occasions, the trials in each set being continued until an event E has occurred exactly r times. For the ith set, the probability of occurrence of E at each trial is θ_i, and the total number of trials is n_i ($i = 1, 2, \ldots, k$). The different sets of trials are mutually independent. Show that the statistic

$$\sum_{i=1}^{k} \left\{ n_i \log\left[\frac{\bar{n}}{n_i}\right] + (n_i - r) \log\left[\frac{n_i - r}{\bar{n} - r}\right] \right\},$$

where $\bar{n} = \sum n_i / k$, may be used to provide a test of the hypothesis that $\theta_1 = \theta_2 = \ldots = \theta_k$ against the alternative that not all the θ_is are equal, and that this test is approximately similar when r is large. (*Camb. Dip.*)

7.6 If, in example 7.5, the principle underlying the chi-squared test is used to test the hypothesis that $\theta_1 = \theta_2 = \ldots = \theta_k$, show that the test statistic which emerges is

$$\frac{r}{\bar{n}(\bar{n}-r)} \sum_{i=1}^{k} (n_i - \bar{n})^2.$$

7.7 Let $r_1, r_2, \ldots, r_k$ be the numbers of occurrences of an event E in k sets of n independent trials. Derive the chi-squared statistic for testing the hypothesis that the probability of occurrence of E at each trial is the same, against the alternative that this probability is constant for any set but may vary from set to set. Verify the theory of the chi-squared test for this example.

7.8 Derive the chi-squared statistic for testing the hypothesis of no association in a contingency table (see section 7.1.3).

7.9 Let $x_1, x_2, \ldots, x_n$ be a large random sample from an $N(\mu, \sigma^2)$ distribution where μ and σ^2 are unknown. Verify the theory underlying the Wald test by considering the problem of testing the hypothesis that $\mu = \sigma^2$.

8 Sequential Tests

The methods of testing hypotheses which we have discussed in chapter 7 are appropriate for answering the question, 'Do we have sufficient evidence to conclude that such-and-such a hypothesis is false?' That is to say, while the Neyman–Pearson theory formulates the problem underlying these methods as one of deciding between two hypotheses, we do not really apply these methods to reach one of the two conclusions 'ω is true' or '$\Theta - \omega$ is true'. Their application leads to one of the slightly weaker conclusions 'we have sufficient evidence to reject ω' or 'we do not have sufficient evidence to reject ω'. The latter conclusion does not imply a conviction that ω is true, for it may well be that the power of the test we have used is small, that is, that the minimum probability of observing a point in the accept-ω region of the test is considerable when $\Theta - \omega$ is true; and this not because we have used an inefficient method, but because the observation we have made is not all that informative. Now in some circumstances we do wish to make a definite decision regarding which of two hypotheses is true, and this may be interpreted as a desire to set a lower limit on the power of a test as well as to control its size, that is, to control both error probabilities.

As we have already seen it is not possible to control both error probability functions arbitrarily with a given experimental design. If therefore we wish such control it is necessary to plan ahead and to choose a design which is informative enough to admit the desired control. To take a somewhat trivial illustration, suppose that we are interested in the probability θ of success in identical independent trials; and that θ is known to have one of the two values θ_1, θ_2. As a result of observation of the results of a number of trials, we wish to decide which of the two hypotheses $\{\theta_1\}$ and $\{\theta_2\}$ is true and we further wish to be fairly confident that the decision we reach, whatever this may be, is correct. This implies that we wish to ensure, in Neyman–Pearson terms, that the probability of each of the possible errors we may make is no more than some preassigned small number, say 0·01. With these requirements there is little point in arbitrarily saying that we will observe the results of six trials and base our decision on this observation. For it may well be not informative enough to enable us to make our decision with the required degree of confidence. For instance suppose that $\theta_1 = 0\cdot 4$ and $\theta_2 = 0\cdot 6$. It is clear on intuitive grounds, as well as a result of the fundamental lemma, that our best possible choice of region for accepting $\{\theta_2\}$ is of the form 'number of successes greater than or equal to k'. If we observe the results of only six trials, then, to make the probability of deciding $\{\theta_2\}$ when $\{\theta_1\}$ is true less than 0·01, we cannot take k less than 6; but if $k = 6$ then the probability of deciding $\{\theta_1\}$ when $\{\theta_2\}$ is true is approximately 0·95 – considerably more than 0·01! Hence it is necessary to think beforehand about our experimental design, that is, in this case, about the number of trials whose results we propose to observe.

Continuing with this particular illustration we may proceed as follows. For each number n of trials observed we might calculate the power of the most-powerful test of significance level 0·01 of the null-hypothesis $\{\theta_1\}$ against the alternative $\{\theta_2\}$. This power will increase with n. We then choose the smallest value of n for which the power is greater than or equal to 0·99 as the number of trials whose results we are going to observe. Such a design would certainly achieve the desired control of error probabilities.

However when we start thinking in terms of preliminary experimental planning with the control of error probabilities in mind, another possibility arises which we have not as yet considered. If observations can be made sequentially, as in our particular illustration, why should we fix in advance the number of observations to be made? Why not, as each observation is taken, look at the results to date to see if as yet we have enough information to enable us to decide between the two hypotheses concerned? If we have, then we reach the appropriate decision without taking any more observations; if we have not, then we take another observation and go through the same procedure. In many ways this is a far more natural procedure than that of fixing in advance the number of observations to be made. It is more in line with the way that a scientist without formal statistical knowledge would proceed. Without formal analysis, however, the assessment of when sufficient information has been obtained to decide between the two hypotheses must be intuitive, and this can be dangerous. Statistical theory enables us to make this assessment on an objective basis and we shall now discuss the theory of one procedure where the number of observations to be taken is not fixed in advance, a so-called *sequential procedure*.

8.1 Definition of a sequential probability ratio test

Let $\{x_n : n = 1, 2, 3, \ldots\}$ be a stochastic process describing some real system which can be observed. The random variables of this process may be independent and identically distributed, as for instance when the process is describing a sequence of independent trials; they may be independent without being identically distributed, as for example, when $x_n = \alpha + \beta t_n + \varepsilon_n$, where $\{\varepsilon_n : n = 1, 2, \ldots\}$ is a process of independent identically distributed random variables; a model appropriate for describing a real system in which a purely random component is superimposed on a variable increasing linearly with time; or the x_ns may be dependent, as for instance, if the process is describing, say, gross national income over the years. We shall start by supposing that the probabilistic structure of this process is known to be one of two possible structures which are defined by the respective sequences of density functions $\{p_n(x_1, x_2, \ldots, x_n) : n = 1, 2, 3, \ldots\}$ and $\{q_n(x_1, x_2, \ldots, x_n) : n = 1, 2, 3, \ldots\}$. Thus we have two simple hypotheses $\mathscr{P}$ and $\mathscr{Q}$, say, regarding the true probabilistic structure describing the real system which we can observe. Suppose that we can observe this system sequentially and that we wish to do so with a view to deciding between $\mathscr{P}$ and $\mathscr{Q}$; and that further we wish to ensure that both error probabilities are controlled at preassigned levels, say, that $\Pr\{\text{deciding } \mathscr{Q} | \mathscr{P}\} \leq \alpha$ and $\Pr\{\text{deciding } \mathscr{P} | \mathscr{Q}\} \leq \beta$. From a practical point of view our assumption that we have merely to decide between two *simple* hypotheses may appear somewhat limiting, but this is a natural theoretical starting point.

One intuitively appealing way of proceeding is to go on taking observations until the probability, under one of the hypotheses, of the observations made, is large enough relative to their probability under the other. Formally we choose two constants $A < 1$ and $B > 1$. We now make an observation x_1. If

$$A < \frac{q_1(x_1)}{p_1(x_1)} < B,$$

then we conclude that this observation does not provide enough information to enable us to decide between $\mathscr{P}$ and $\mathscr{Q}$; if $q_1(x_1)/p_1(x_1) \leq A$ we decide that $\mathscr{P}$ is true and stop observation; if $q_1(x_1)/p_1(x_1) \geq B$ we decide that $\mathscr{Q}$ is true and stop observation. We continue to take observations in this way until either $\lambda_n = q_n(x_1, x_2, \ldots, x_n)/p_n(x_1, x_2, \ldots, x_n) \leq A$, in which case we decide that $\mathscr{P}$ is true or $\lambda_n \geq B$ in which case we decide that $\mathscr{Q}$ is true.

Such a procedure is called a *sequential probability ratio test*, (s.p.r. test).

8.2 Error probabilities and the constants A and B

The main analytic problem connected with an s.p.r test is that of relating the test constants A and B to its error probabilities. As indicated above, in practice we will be given upper limits to the error probabilities and will wish to choose A and B so that actual error probabilities are within these preassigned limits. Moreover, since taking observations can be costly, we will wish to choose A and B in an economic way, or more formally in such a way that the expected number of observations to be taken before a decision is reached, either when $\mathscr{P}$ is true or when $\mathscr{Q}$ is true, is minimized if this is possible. It can be shown by a rather difficult proof (Lehmann, 1959, p. 98), that among all procedures satisfying

$\Pr\{\text{deciding } \mathscr{Q} | \mathscr{P}\} \leq \alpha$
and $\Pr\{\text{deciding } \mathscr{P} | \mathscr{Q}\} \leq \beta,$

the s.p.r. test for which these probabilities are respectively equal to α and β is most economic in this sense. In general, however, the determination of A and B

to make these error probabilities exactly α and β respectively for the s.p.r. test is extremely difficult. Fortunately it is easy to obtain values of A and B for which the error probabilities are in most cases approximately α and β; and this is done as follows.

Consider the s.p.r. test which uses the numbers A and B. Let its error probabilities be α' and β'. All sets $(x_1, x_2, \ldots, x_n)$ of observations which lead to the decision $\mathcal{Q}$ are at least B times more probable when $\mathcal{Q}$ is true than when $\mathcal{P}$ is true. By 'integrating $q(x_1, x_2, \ldots, x_n)$ over all such sets of observations' we obtain $\Pr\{\text{deciding } \mathcal{Q} | \mathcal{Q}\}$. By integrating $p(x_1, x_2, \ldots, x_n)$ over the same sets, we obtain $\Pr\{\text{deciding } \mathcal{Q} | \mathcal{P}\}$.

Therefore $\quad \Pr\{\text{deciding } \mathcal{Q} | \mathcal{Q}\} \geqslant B \Pr\{\text{deciding } \mathcal{Q} | \mathcal{P}\}$.

Now $\qquad \Pr\{\text{deciding } \mathcal{Q} | \mathcal{P}\} = \alpha'$.

Suppose that the s.p.r. test is such that, when $\mathcal{Q}$ is true, with probability 1, it terminates after a finite number of observations. Then

$$\Pr\{\text{deciding } \mathcal{Q} | \mathcal{Q}\} = 1 - \beta'.$$

Hence $\quad 1 - \beta' \geqslant B\alpha'$,

or $\qquad B \leqslant \dfrac{1-\beta'}{\alpha'}.$

Similarly if we assume that when $\mathcal{P}$ is true the s.p.r. test terminates after a finite number of observations with probability 1, we have

$$A \geqslant \frac{\beta'}{1-\alpha'}.$$

We now examine these inequalities to see how sharp they are. The only reason why the first is an inequality rather than an equality is the fact that the decision '$\mathcal{Q}$ is true' may be reached with $\lambda_n > B$ rather than $\lambda_n = B$; in other words, that the likelihood ratio 'overshoots the boundary' at B. If with probability near 1, this overshoot is small, then B will be approximately equal to $(1-\beta')/\alpha'$. Suppose for instance that the decision '$\mathcal{Q}$ is true' is always reached with $\lambda_n \leqslant B + \delta$. Then we should have

$$B \leqslant \frac{1-\beta'}{\alpha'} \leqslant B + \delta.$$

In most practical cases this tends to happen; the likelihood ratio 'creeps up' on the boundary at B and when it crosses it, the overshoot is small. This is because usually each successive observation provides only a little information for discriminating between $\mathcal{P}$ and $\mathcal{Q}$. Of course it is possible to construct examples where the likelihood ratio suddenly leaps far over the boundary and the reader may find it instructive to do so. However, more commonly in practice, the inequality $B \leqslant (1-\beta')/\alpha'$ is indeed very nearly an equality;

and obviously for the same reason, the same applies to the inequality $A \geq \beta'/(1-\alpha')$. This suggests that if we wish to construct a s.p.r. test with error probabilities approximately α and β then we should take $A = \beta/(1-\alpha)$ and $B = (1-\beta)/\alpha$. That this is not merely a speculative result has been confirmed in practice by extensive numerical calculations in particular cases, so that it is now a well-established practical method.

We assumed, in establishing these inequalities that the s.p.r. test concerned terminates with probability 1, that is, that the probability is zero that we go on taking observations for ever without reaching a decision. It is not difficult to prove that this assumption is valid if the sequence (x_n) is one of independent identically distributed random variables under either hypothesis,

so that $\quad p_n(x_1, x_2, \ldots, x_n) = \prod_{i=1}^{n} p(x_i)$

and $\quad q_n(x_1, x_2, \ldots, x_n) = \prod_{i=1}^{n} q(x_i),\quad$ say,

where the distributions defined by the density functions p and q are essentially different. This we leave to the reader as an example on the strong law of large numbers, an example which uses also a result which we established in section 4.4. While the case of independent, identically distributed random variables is one of great practical importance, the assumption of termination with probability 1 has considerably wider validity. It is difficult to introduce an interesting theorem of sufficient generality to cover this point, though it is easy enough to see why the assumption can fail. If, eventually, successive observations provide little additional information for discriminating between $\mathscr{P}$ and $\mathscr{Q}$ then the likelihood ratio λ_n can stay more or less fixed for ever and fail to reach one of the boundaries A and B. This can happen for instance, if, for some integer m the conditional distributions of $x_{m+1}, x_{m+2}, \ldots$ given $x_1, x_2, \ldots, x_m$ are the same under both $\mathscr{P}$ and $\mathscr{Q}$. The observation of $x_{m+1}, x_{m+2}, \ldots$ gives no discrimatory information additional to that given by $x_1, x_2, \ldots, x_m$: in other words $\lambda_m = \lambda_{m+1} = \lambda_{m+2} = \ldots$, so that if we have not reached a boundary by the mth observation we can never do so. Usually this danger will be apparent in practice, but it is worth remembering whenever an s.p.r. test is applied in a situation where successive observations are dependent.

8.3 Graphical procedure for an s.p.r. test

It sometimes happens that the s.p.r. test can be reduced to a graphical procedure which is extremely easy to apply. Consider, for example, the problem of determining whether the probability of success in independent identical trials is θ_1 (the hypothesis $\mathscr{P}$) or θ_2 (the hypothesis $\mathscr{Q}$). We may take the sequence (x_n) to be one of independent random variables each of which takes the values 0 and 1.

Let $X_n = \sum_{i=1}^{n} x_i$ $(n = 1, 2, \ldots)$.

Then $\lambda_n = \dfrac{\theta_2^{X_n}(1-\theta_2)^{n-X_n}}{\theta_1^{X_n}(1-\theta_1)^{n-X_n}}$.

Now $A < \lambda_n < B$ iff $\log A < \log \lambda_n < \log B$,

i.e. iff $\log A < X_n \log \dfrac{\theta_2}{\theta_1} + (n - X_n) \log \dfrac{1-\theta_2}{1-\theta_1} < \log B$,

or $\log A - n \log \dfrac{1-\theta_2}{1-\theta_1} < X_n \log \dfrac{\theta_2(1-\theta_1)}{\theta_1(1-\theta_2)} < \log B - n \log \dfrac{1-\theta_2}{1-\theta_1}$.

Suppose that $\theta_2 > \theta_1$.

Then $\dfrac{\theta_2(1-\theta_1)}{\theta_1(1-\theta_2)} > 1$ and $\log \dfrac{\theta_2(1-\theta_1)}{\theta_1(1-\theta_2)} > 0$.

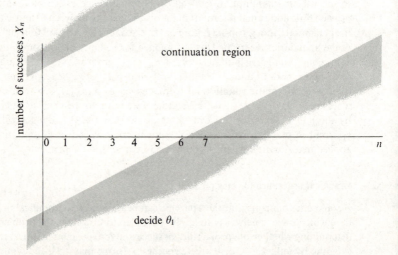

Figure 5 Graphical procedure for an s.p.r. test to decide between two possible probabilities for each of a sequence of independent trials

In this case the s.p.r. test using the constants A and B continues so long as
$a+cn < X_n < b+cn$,

where $\quad a = \dfrac{\log A}{\log [\theta_2(1-\theta_1)/\{\theta_1(1-\theta_2)\}]} \quad b = \dfrac{\log B}{\log [\theta_2(1-\theta_1)/\{\theta_1(1-\theta_2)\}]}$

and $\quad c = -\dfrac{\log [(1-\theta_2)/(1-\theta_1)]}{\log [\theta_2(1-\theta_1)/\{\theta_1(1-\theta_2)\}]}.$

The test terminates with the decision '$\mathscr{P}$ is true' if these inequalities are first violated by $X_n \leq a+cn$, and with the decision '$\mathscr{Q}$ is true' if the first violation is by $X_n \geq b+cn$. By plotting X_n against n in a graph containing the lines $X_n = a+cn$ and $X_n = b+cn$ we arrive at the simple graphical procedure promised. This is illustrated in Figure 5. Note that X_n is simply the total number of successes in the first n trials and that the common slope and relative positions of the two lines follow from our assumption that $\theta_2 > \theta_1$. The change to be made if $\theta_2 < \theta_1$ is bound up with the fact that then

$$\log \left\{ \dfrac{\theta_2(1-\theta_1)}{\theta_1(1-\theta_2)} \right\} < 0$$

so that inequality signs have to be changed on division by this quantity.

Finally we note that if upper limits α and β to error probabilities are pre-assigned rather than the numbers A and B, then we obtain a test whose error probabilities are approximately α and β by taking

$$A = \dfrac{\beta}{1-\alpha} \quad \text{and} \quad B = \dfrac{1-\beta}{\alpha}.$$

8.4 Composite hypotheses

It is seldom in practice that one is faced with a problem of testing one simple hypothesis against another. More often than not each hypothesis is composite and, not surprisingly, the problem of sequential tests between two composite hypotheses is more severe than that which we have been discussing. The formal extension of the latter problem is this: the probabilistic structure of a stochastic process (x_n) is labelled by a parameter θ ranging over the parameter space Θ, and we wish to establish a procedure for deciding between two composite hypotheses ω and $\Theta - \omega$, this procedure to have the properties

$$P_\theta\{\text{deciding } \Theta - \omega\} \leq \alpha \quad \text{for all } \theta \in \omega$$
$$\text{and} \quad P_\theta\{\text{deciding } \omega\} \leq \beta \quad \text{for all } \theta \in \Theta - \omega.$$

That is, we wish a test which controls both error probabilities.

It may well be that this problem is insoluble. Consider, for instance, the problem of deciding whether the probability θ of success in a sequence of independent identical trials is such that $\theta \leq \theta_0$ or that $\theta > \theta_0$. If both α and β

are less than a half, as of course they usually will be, then there is certainly no fixed sample size procedure which controls the probabilities of error as we wish. Such a procedure, based on the results of N trials, divides the possible results into two sets, an 'accept $\theta \leqslant \theta_0$ set', A_0, say and its complement A'_0, the 'accept $\theta > \theta_0$ set'; and, to meet our demands, these sets must be such that

$$P_\theta(A'_0) \leqslant \alpha \quad \text{for all } \theta \leqslant \theta_0$$
and $\quad P_\theta(A_0) \leqslant \beta \quad \text{for all } \theta > \theta_0$.

In particular $P_{\theta_0}(A'_0) \leqslant \alpha$ and so $P_{\theta_0}(A_0) \geqslant 1-\alpha > \frac{1}{2}$, if $\alpha < \frac{1}{2}$. Now $P_\theta(A_0)$ is a continuous function of θ and so there exists a $\theta > \theta_0$ such that $P_\theta(A_0) > \frac{1}{2}$. This conflicts with the demand that $P_\theta(A_0) \leqslant \beta$ for all $\theta > \theta_0$, if $\beta < \frac{1}{2}$. Thus, as we have said, there is no fixed sample size procedure which meets our requirements. Intuitively it is fairly clear that there is no sequential procedure, terminating after a finite number of observations with probability 1, which satisfies these demands.

The general difficulty is now apparent. If, for each set A_0 in the sample space, $P_\theta(A_0)$ is continuous in the parameter θ labelling the family of possible distributions and if the sets ω and $\Theta - \omega$ defining the hypotheses under test have a common boundary, it is impossible to construct a practical test for deciding between ω and $\Theta - \omega$ with arbitrary preassigned upper limits to both error probabilities. Faced with this impossibility we must weaken our demands on a procedure for deciding between two hypotheses of the nature stated.

We illustrate one possible line of argument by referring again to the problem of testing whether $\theta \leqslant \theta_0$ or $\theta > \theta_0$, θ being the probability of success in each of a sequence of independent trials. If we look at the consequences of wrong decisions it may be that we can find two numbers $\theta_1 < \theta_0$ and $\theta_2 > \theta_0$ such that the error of deciding that $\theta > \theta_0$, when in fact $\theta_1 < \theta \leqslant \theta_0$, is not all that serious, and the error of deciding that $\theta \leqslant \theta_0$, when in fact $\theta_0 < \theta < \theta_2$, is not serious either. In other words, if $\theta_1 < \theta < \theta_2$ it is immaterial whether we decide that $\theta \leqslant \theta_0$ or that $\theta > \theta_0$. Then we may set out to construct a test between the hypotheses $\omega_1 = \{\theta : \theta \leqslant \theta_1\}$ and $\omega_2 = \{\theta : \theta \geqslant \theta_2\}$ with preassigned upper limits to the probabilities of error. In this way we control the probabilities of 'serious' errors and allow less serious errors to take care of themselves. This is in line with the Neyman–Pearson analysis in which Type I error is regarded as more serious than the Type II and in a sense it carries this analysis one stage further. (It may appear to the reader that this further stage still leaves the analysis incomplete; we shall return to this point later.)

8.5 Monotone likelihood ratio and the s.p.r. test

8.5.1 Suppose that in the meantime we adopt this point of view. Then in the particular case of testing between ω_1 and ω_2 above, it is plausible that the s.p.r. test with error probabilities α and β between the *simple* hypotheses $\{\theta_1\}$ and

$\{\theta_2\}$ has the properties:

$$P_\theta\{\text{deciding } \theta_2\} \leq \alpha \quad \text{for all } \theta \leq \theta_1$$
$$\text{and} \quad P_\theta\{\text{deciding } \theta_1\} \leq \beta \quad \text{for all } \theta \geq \theta_2.$$

8.1

This is true, as well as plausible, and the general result of which it is a particular case is as follows.

Suppose that the stochastic process (x_n) is one of independent identically distributed random variables whose common probability density function p_θ depends on a real parameter θ; and that the family $\{p_\theta\}$ of densities has *monotone likelihood ratio*, that is, for $\theta' > \theta$ the ratio $p_{\theta'}(x_i)/p_\theta(x_i)$ is a non-decreasing function of some statistic $t(x_i)$. Then the s.p.r. test between $\{\theta_1\}$ and $\{\theta_2\}$ ($\theta_1 < \theta_2$) with error probabilities α and β has the properties **8.1** above. We can therefore use this procedure in the obvious way as a sequential procedure for deciding between the hypotheses

$$\omega_1 = \{\theta : \theta \leq \theta_1\}$$
$$\text{and} \quad \omega_2 = \{\theta : \theta \geq \theta_2\},$$

and thereby satisfy preassigned upper limits to the probabilities of wrong decisions.

This result has been proved by Lehmann (1959), p. 101, and we refer the reader to his proof.

In very particular circumstances we now have a practical sequential procedure for deciding between two composite hypotheses, ω_1 and ω_2. Before going on to discuss what can be done when these circumstances do not obtain, we shall consider some problems concerned with the particular case. To be specific, (x_n) is a process of independent identically distributed random variables; the family of possible distributions of each x_i is labelled by a real parameter θ, and defined by a family $\{p_\theta\}$ of density functions with monotone likelihood ratio; we continue to observe the stochastic process (x_n) so long as

$$A < \prod_{i=1}^{n} \frac{p_{\theta_2}(x_i)}{p_{\theta_1}(x_i)} < B;$$

if these inequalities are first violated by the central ratio taking a value less than or equal to A, we decide that $\theta \leq \theta_1$; if by its taking a value greater than or equal to B, we decide that $\theta \geq \theta_2$. This procedure has the properties that

$$P_\theta\{\text{deciding } \theta \geq \theta_2\} \leq P_{\theta_1}\{\text{deciding } \theta \geq \theta_2\} \quad \text{for all } \theta \leq \theta_1,$$
$$\text{and} \quad P_\theta\{\text{deciding } \theta \leq \theta_1\} \leq P_{\theta_2}\{\text{deciding } \theta \leq \theta_1\} \quad \text{for all } \theta \geq \theta_2.$$

As we have seen, by taking $A = \beta/(1-\alpha)$, and $B = (1-\beta)/\alpha$, we can ensure that, approximately,

$$P_{\theta_1}\{\text{deciding } \theta \geq \theta_2\} = \alpha$$
$$\text{and} \quad P_{\theta_2}\{\text{deciding } \theta \leq \theta_1\} = \beta.$$

The number n of observations required by this procedure before a decision is reached is a random variable whose distribution depends on the parameter θ. In particular the expected number $E_\theta(n)$ of observations required is a function of θ which may be of interest to the experimenter; in addition he may be interested in knowing the exact value of $P_\theta\{\text{deciding } \theta \geq \theta_2\}$ for various values of θ. Answers to these questions can be provided by exploiting the theory of random walks with two absorbing barriers and in particular by using Wald's identity. (Familiarity with this theory is assumed in what follows. An excellent account of it is given by Cox and Miller, 1965.) To see this, consider the sequential procedure adopted. Observation continues so long as

$$A < \prod_{i=1}^{n} \frac{p_{\theta_2}(x_i)}{p_{\theta_1}(x_i)} < B,$$

or, equivalently, $\quad \log A < \sum_{i=1}^{n} y_i < \log B,$

where $\quad y_i = \log \dfrac{p_{\theta_2}(x_i)}{p_{\theta_1}(x_i)}.$

Since $x_1, x_2, \ldots$ are independent and identically distributed, so are $y_1, y_2, \ldots$. Therefore we are dealing essentially with a random walk, starting at 0, and having absorbing barriers at $a = \log A < 0$ and $b = \log B > 0$. Absorption at a means that the decision $\theta \leq \theta_1$ is reached; absorption at b implies the decision $\theta \geq \theta_2$. The theory of random walks therefore becomes relevant.

8.5.2 Let n denote the number of observations required to reach a decision, and let $\phi(z)$ be the moment-generating function of each y_i. (This depends on the true parameter θ.) Wald's identity states that

$$E\{[\phi(z)]^{-n} e^{zY_n}\} \equiv 1,$$

where $Y_n = y_1 + y_2 + \ldots + y_n$. This identity is valid for all values of z except possibly that at which the convex function $\phi(z)$ takes its minimum value. In particular it is valid for the two values of z, $z = 0$ and $z = z_1$, say, for which $\phi(z) = 1$, so that we have

$$E(e^{z_1 Y_n}) = 1.$$

This may be written

$$P_a E_a(e^{z_1 Y_n}) + P_b E_b(e^{z_1 Y_n}) = 1,$$

where P_a denotes the probability of absorption at a (the decision $\theta \leq \theta_1$) and E_a denotes expectation conditional on absorption at a, and similarly for P_b and E_b. Now given absorption at a, Y_n takes values in the range $Y_n \leq a$. If we neglect overshoot of the barrier, then, given absorption at a, $Y_n = a$; and similarly, neglecting overshoot, given absorption at b, $Y_n = b$. As we have seen, for the s.p.r. test, overshoot is probably very small. Therefore, approximately,

$$P_a e^{z_1 a} + P_b e^{z_1 b} = 1.$$

Moreover, absorption at one or other barrier is certain so that

$$P_a + P_b = 1.$$

These equations may be solved to give approximations to P_a and P_b.

Differentiation of Wald's identity with respect to z, and substitution of the value 0 for z, yields the result that

$$E(n) = \frac{E(Y_n)}{E(\text{each } y_i)},$$

provided $E(\text{each } y_i) \neq 0$. If again we neglect overshoot of the barriers, Y_n takes only two values, a with probability P_a, and b with probability P_b, and so

$$E(n) \simeq \frac{aP_a + bP_b}{E(\text{each } y_i)}.$$

If $E(\text{each } y_i) = 0$, then it can be shown by differentiating Wald's identity twice with respect to z, and substituting 0 for z, that

$$E(n) = \frac{E(Y_n^2)}{E\{(\text{each } y_i)^2\}}.$$

Now in this case – when $E(y_i) = 0$ – the equation $\phi(z) = 1$ has the repeated root $z = 1$, so that the above method for determining approximations to P_a and P_b breaks down. Further analysis shows that P_a is then approximately $b/(b-a)$, while P_b is approximately $-a/(b-a)$, so that

$$E(Y_n^2) \simeq \frac{a^2 b}{b-a} - \frac{b^2 a}{b-a} = -ab$$

and $\quad E(n) \simeq \dfrac{-ab}{E\{(\text{each } y_i)^2\}}.$

8.5.3 Example

From observations of the results of independent identical trials it is desired to test sequentially whether $\theta \leqslant \theta_0$ or $\theta > \theta_0$, θ being the common probability of success. The test is required to have the properties that

$$P_\theta\{\text{deciding } \theta > \theta_0\} \leqslant \alpha \quad \text{for all } \theta \leqslant \theta_1$$
$$\text{and} \quad P_\theta\{\text{deciding } \theta \leqslant \theta_0\} \leqslant \beta \quad \text{for all } \theta \geqslant \theta_2,$$

θ_1 and θ_2 being numbers such that $0 < \theta_1 < \theta_2 < 1$. Accordingly boundaries on the number of successes are determined by the s.p.r. test between $\{\theta_1\}$ and $\{\theta_2\}$ having approximate error probabilities α and β. The experimenter wishes to know 'how many observations will be required by this procedure and what its error probabilities are'. What answers can be given to these questions?

The procedure used may be described as follows in order to exhibit the relevance of the random walk theory just stated. If the ith trial results in success, set $y_i = \log(\theta_2/\theta_1)$; if in failure, set $y_i = \log\{(1-\theta_2)/(1-\theta_1)\}$. After m trials ($m = 1, 2, \ldots$) calculate

$$Y_m = y_1 + y_2 + \ldots + y_m.$$

So long as $\quad a = \log \dfrac{\beta}{1-\alpha} < Y_m < \log \dfrac{1-\beta}{\alpha} = b,$

carry out another trial. Suppose that these inequalities are first violated when $m = n$. Then if $Y_n \leqslant a$, decide that $\theta \leqslant \theta_0$; if $Y_n \geqslant b$ decide that $\theta > \theta_0$. This actual procedure, of course, does not depend on the true value of the probability of success, but its properties do so depend. Suppose that the true probability of success is θ. Then the sequence (Y_m), with Y_0 defined to be 0, is a random walk starting at 0, having absorbing barriers at $a < 0$ and $b > 0$, and each of the increment random variables, y_i, takes the values

$$\log \frac{\theta_2}{\theta_1} \quad \text{with probability } \theta$$

and $\quad \log \dfrac{1-\theta_2}{1-\theta_1} \quad$ with probability $1-\theta$.

It is in the probabilities with which each increment takes its two possible values that the true probability of success enters the picture. When this probability is θ, the moment-generating function $\phi(z)$ of each y_i is

$$\phi(z) = \theta \exp\left[z \log \frac{\theta_2}{\theta_1}\right] + (1-\theta) \exp\left[z \log \frac{1-\theta_2}{1-\theta_1}\right]$$

$$= \theta \left[\frac{\theta_2}{\theta_1}\right]^z + (1-\theta)\left[\frac{1-\theta_2}{1-\theta_1}\right]^z.$$

For any particular value of θ we may solve numerically the equation $\phi(z) = 1$ in order to determine the value of z_1 corresponding to this θ, and then the general formulae of this section with $a = \log \beta/(1-\alpha)$ and $b = \log(1-\beta)/\alpha$ may be used to calculate the probabilities of the two possible decisions associated with this value of θ; and also $E_\theta(n)$.

The detailed calculation involved here is often more than is necessary to answer the experimenter's questions for, as often as not, he is interested only in obtaining a rough idea of, for example, $E_\theta(n)$. This can be gained by determining $E_\theta(n)$ for several particular values of θ for which the calculation is easy: these values are $\theta = 0, 1, \theta_1, \theta_2$ and that value of θ for which $E(y_i) = 0$,

namely $\quad \theta = \dfrac{-\log\{(1-\theta_2)/(1-\theta_1)\}}{\log[\theta_2(1-\theta_1)/\{\theta_1(1-\theta_2)\}]}.$

Obviously $\quad E_0(n) = \left[\dfrac{a}{\log\{(1-\theta_2)/(1-\theta_1)\}}\right] + 1$

and $\quad E_1(n) = \left[\dfrac{b}{\log(\theta_2/\theta_1)}\right] + 1,$

where $[c]$ denotes the greatest integer smaller than c, for in either case the random walk is no longer random; it is deterministic, rather.

If $\theta = \theta_1$, then $z_1 = 1$, and we have

$$P_a e^a + P_b e^b \simeq 1$$
$$P_a + P_b = 1$$

so that $\quad P_a \simeq \dfrac{e^b - 1}{e^b - e^a} = \dfrac{\{(1-\beta)/\alpha\} - 1}{\{(1-\beta)/\alpha\} - \beta/(1-\alpha)} = 1 - \alpha,$

a result which of course is obvious but whose derivation by this method is of some interest.

Hence $\quad E_{\theta_1}(n) \simeq \dfrac{(1-\alpha)\log\{\beta/(1-\alpha)\} + \alpha\log\{(1-\beta)/\alpha\}}{\theta_1 \log(\theta_2/\theta_1) + (1-\theta_1)\log\{(1-\theta_2)/(1-\theta_1)\}}.$

Similarly $\quad E_{\theta_2}(n) \simeq \dfrac{\beta\log\{\beta/(1-\alpha)\} + (1-\beta)\log\{(1-\beta)/\alpha\}}{\theta_2 \log(\theta_2/\theta_1) + (1-\theta_2)\log\{(1-\theta_2)/(1-\theta_1)\}}.$

Finally, when $\quad \theta = \dfrac{-\log\{(1-\theta_2)/(1-\theta_1)\}}{\log\{\theta_2(1-\theta_1)/\theta_1(1-\theta_2)\}} = \theta^*, \quad$ say.

$E_{\theta^*}(n)$
$\simeq \dfrac{-\log\{\beta(1-\alpha)^{-1}\}\log\{(1-\beta)\alpha^{-1}\}/[\log\{(1-\beta)\alpha^{-1}\} - \log\{\beta(1-\alpha)^{-1}\}]}{\theta^*\{\log(\theta_2/\theta_1)\}^2 + (1-\theta^*)[\log\{(1-\theta_2)/(1-\theta_1)\}]^2}.$

These five values of the function $E_\theta(n)$ of θ are enough to enable us to sketch the graph of this function, which has a maximum at θ^*. Graphs of error probability functions can be drawn similarly and the experimenter's questions are answered by these graphs.

8.5.4 The procedure which we have been discussing is applicable generally in the circumstances described in section 8.5.2 and so we have a method for dealing with an important class of practical problems. However, looked at from the general point of view of sequential tests between two composite hypotheses, the circumstances (a single unknown parameter, monotone likelihood ratio, etc.) are very particular indeed, and the general problem remains. That is, we have no *general* method of constructing tests between composite hypotheses. Moreover there are difficult problems involved in discussing properties, such as error probability functions and expected number functions of particular sequential tests when we cannot appeal to random walk theory. Certain additional more complex practical problems can be reduced to the monotone likelihood-ratio form by invoking the principle of invariance, see Cox (1952);

and Wald (1947) suggests a method of tackling the general problem. (Incidentally this book by Wald contains an excellent discussion of the s.p.r. test.) However further development of the theory of sequential tests is contained within decision theory which we shall discuss in chapter 11.

Finally it should be noted that we have not demonstrated that the particular procedure described in section 8.5.1 is optimal in any way. It is a procedure which happens, in special circumstances, to control error probabilities as we wish to control them, but we have not shown, for example, that among all procedures which control error probabilities in this way, this particular one uniformly minimizes $E_\theta(n)$, regarded as a function of θ. Indeed it is known (see Lehmann, 1959, p. 102) that it does not do so in general. Alternative procedures are available in the special circumstances of monotone likelihood ratio etc., which place an upper limit on the number of observations to be made. These are particularly useful in medical trials and for an account of these so-called *closed sequential designs*, the reader is referred to Armitage (1960). Another useful book on sequential tests is by Wetherill (1966).

Examples

8.1 Observations $x_1, x_2, \ldots, x_n$ form a random sample from a normal distribution with variance 1 and unknown mean θ. What is the minimum value of n for which it is possible to find a region R in the sample space such that

$$P(R|\theta = 0) = 0.05$$
and $$P(R|\theta = \tfrac{1}{2}) = 0.95?$$

8.2 In a sampling scheme to control the quality of bricks, random samples of size n were taken from each load, which was accepted as satisfactory or rejected according as $\bar{x} \leq k$ or $\bar{x} > k$, where $\bar{x}$ is the mean specific gravity of the bricks in the sample, and k is a constant. It was desired that if θ, the mean specific gravity of a load, exceeded 2·42 the probability of its rejection should be at least 0·99, while if $\theta \leq 2\cdot38$, the probability of its acceptance should be at least 0·95. The standard deviation of specific gravity of the individual bricks in a load could be taken as 0·03. Assuming that n is small compared with the number of bricks in a load and that the distribution of $\bar{x}$ is normal, find the least value of n satisfying these requirements.

8.3 Let $x_1, x_2, \ldots$ be a sequence of independent, identically distributed random variables and let H_0 and H_1 be two simple hypotheses concerning the distribution of each x_i. In each of the following cases, construct a simple graphical procedure for applying a sequential probability ratio test between H_0 and H_1 having both error probabilities approximately equal to 0·05.
(a) H_0: each x_i is $N(0, 1)$; H_1: each x_i is $N(2, 1)$.
(b) H_0: each x_i has density e^{-x} $(x > 0)$; H_1: each x_i has density $2e^{-2x}$ $(x > 0)$.

(c) H_0: each x_i is Poisson with mean $\frac{1}{2}$; H_1: each x_i is Poisson with mean 1.

8.4 In a factory components are manufactured in two qualities, 1 and 2, and the lifetimes in hours of these two types are distributed with densities xe^{-x} and $4xe^{-2x}$ ($x > 0$, respectively). A large batch of components taken from store is found to be unlabelled, though it is known to consist of components of the same quality. Since differences in quality cannot be detected visually, it is proposed to measure lifetimes of components sequentially until a decision on quality can be reached.

Suppose that there are preassigned upper limits α and β to the probabilities of wrong decisions. Derive an appropriate procedure for the manufacturer and show how it may be carried out graphically.

If the manufacturer asks how long he is likely to have to wait before a decision is reached, what answer would you give?

Suppose that the components in the batch tested do not in fact belong to either of types 1 and 2 and that their lifetimes in hours are distributed with density $\lambda^2 x e^{-\lambda x}$. Let P_λ be the probability of deciding, according to the above procedure with $\alpha = \beta = 0.01$, that they are of quality 2. Sketch the graph of P_λ. Also sketch the graph of E_λ, the expected number of observations required for a decision.

8.5 Independent observations $x_1, x_2, \ldots$ are obtained sequentially. Each is normally distributed with mean μ and variance σ^2. Two sequential tests of the null-hypothesis $\sigma = \sigma_0$ against the alternative $\sigma = \lambda\sigma_0$, where λ is a given constant less than 1, are proposed, μ being unknown. The first takes the observations in pairs and uses $x_{2i} - x_{2i-1}$, a decision being reached after each pair. The second uses the observations in the form $x_2 - x_1, 2x_3 - (x_1 + x_2), \ldots$ and generally $nx_{n+1} - \sum_{i=1}^{n} x_i$. Describe how each of these can be constructed to give preassigned probabilities of error under the two hypotheses. Is one of these procedures better than the other? (See also the sequential F-test in Wetherill, 1966.)

8.6 Independent observations $x_1, x_2, \ldots$ are made sequentially on the DNA content of cells. There are two possibilities. Either the cells are all of one type in which case each x_i is normally distributed with known mean μ and known variance σ^2; or there are two types of cell mixed in proportion 9:1, in which case the distribution of each x_i is a mixture of an $N(\mu, \sigma^2)$ and an $N(2\mu, 2\sigma^2)$ distribution, the mixing probabilities being respectively 0.9 and 0.1. If a sequential probability ratio test with error probabilities of 0.01 is conducted to decide between these alternatives, and an experimenter wishes a preliminary estimate of the number of observations he will require, what answer would you give?

8.7 $(x_n; n = 1, 2, \ldots)$ is a two-state homogeneous Markov process with states 1 and -1. Transition probabilities are either

(a) $P(x_n = 1 | x_{n-1} = 1) = \tfrac{3}{4}$ and $P(x_n = 1 | x_{n-1} = -1) = \tfrac{1}{4}$

or (b) $P(x_n = 1 | x_{n-1} = 1) = \tfrac{1}{4}$ and $P(x_n = 1 | x_{n-1} = -1) = \tfrac{3}{4}$.

Let r_n be the number of changes of state occurring in the first n observations of this process. Show that it is possible to choose a and b in such a way that a sequential procedure for deciding between (a) and (b) with continuation region

$$a - \tfrac{1}{2}n < r_n < b - \tfrac{1}{2}n$$

has error probabilities approximately α and β, and determine these values when the distribution of x_1 is the same under both hypotheses.

9 Non-Parametric Methods

In previous chapters we have been concerned primarily with parametric problems, that is, with real situations where the appropriate probability distribution on the sample space of the mathematical model is known apart from the values of a finite number of unknown parameters. We have attempted to state criteria which characterise a good solution of the problem in hand and more often than not this attempt has failed in the sense that no solution exists which satisfies the demands of our criteria. Nevertheless the attempt is not completely in vain, for it establishes a framework within which we can discuss properties of general methods introduced as intuitively reasonable, methods such as maximum-likelihood estimation, the likelihood-ratio test and the sequential probability ratio test; and within which we can compare properties of different solutions to the same problem.

Non-parametric theory is concerned with problems involving larger families of possible distributions, families which cannot be labelled by a finite-dimensional vector-valued parameter. Often there are two possible approaches to a given problem in inference, the parametric and non-parametric. The former involves stronger assumptions than the latter about the family of possible distributions on the sample space, assumptions which may not be verifiable, and to this extent the non-parametric approach is more realistic. To illustrate this, suppose that we have data consisting of measurements of some characteristic on a random sample of m control units and on an independent random sample of n units which have been subjected to a treatment. Thus we have a random sample, $x_1, x_2, \ldots, x_m$, say, from one distribution and an independent random sample $y_1, y_2, \ldots, y_n$, say, from another. Let these distributions have distribution functions F and G respectively. We may be prepared to assume that the only possible effect of the treatment is to 'shift the mean', that is, that

$$G(z) = F(z-\mu),$$

where μ is an unknown constant; and then we are interested in estimating μ from the data. One approach to this problem of estimation is to assume normality of the underlying distributions. The problem then becomes parametric in character and we may obtain a 'best' estimate of μ by the methods of chapter 2. We may also derive a confidence interval for μ; but our confidence in this interval will depend on our confidence in the assumption of normality which we have made. If there is good reason for this assumption, we will be quite happy about the confidence interval. However if the assumption has been made on the grounds of expediency, then we might well require a 'robustness' study, that is a study of the effect on the confidence interval of departures from normality of the underlying distributions, in order to bolster up our faith in the proposed interval.

The alternative approach is to make far weaker assumptions about the nature of the underlying distributions. For instance, in the above illustration we might assume merely that F is continuous, a very weak assumption about which we could completely confident in a given situation. The problem is now non-parametric because there is a distribution on the sample space of our model corresponding to each distribution on the line with continuous distribution function, and this family cannot be labelled by a finite-dimensional vector-valued parameter – it is far too large. The labelling parameter θ for this family now takes the form $\theta = (F, \mu)$ and F ranges over the space of continuous distribution functions on the line, while μ ranges over the real numbers. Of course, despite the size of the family of possible distributions, we may still attempt to find a $100(1-\alpha)$ per cent confidence interval for μ, that is, an interval $S(x, y)$ (where $(x, y) = (x_1, x_2, \ldots, x_m, y_1, y_2, \ldots, y_n)$), defined for each (x, y) and having the property that

$$P_\theta\{(x, y); \mu \in S(x, y)\} = 1 - \alpha, \quad \text{for all } \theta.$$

Naturally, the larger the family of possible distributions, the more severe this problem becomes. So, not surprisingly, we should be content initially to find any solution to it without worrying unduly about wether the solution is optimum relative to some stated criteria.

This illustrates a general point about non-parametric theory. We did not have great success in stating criteria which led to unique optimum solutions to parametric problems; we might expect even less success in such an approach to non-parametric problems. And this is reflected in the historical development of non-parametric theory. Methods were suggested for particular problems and only relatively recently have optimum properties of these methods been investigated. In this chapter we shall discuss a few of these methods, without saying much about their properties, though some of the notions introduced in earlier chapters such as the size and power of tests provide a useful structure for thinking about the methods.

9.1 The Kolmogorov–Smirnov test

The first non-parametric problem which we shall consider is as follows. Given a random sample $x = (x_1, x_2, \ldots, x_n)$ from a continuous distribution on the line, it is desired to test the *simple* hypothesis that the distribution function is the specified one, F. We shall refer to this as the null-hypothesis to indicate that our conclusion will be either, 'There is enough evidence to reject this hypothesis' or, 'There is not enough evidence to reject it.'

The Kolmogorov–Smirnov test is a very natural one and is defined thus:

Let the ordered sample be $x_{(1)} < x_{(2)} < \ldots < x_{(n)}$ and let $F_n(z)$ be the *sample* or *empirical* distribution function defined by

$$F_n(z) = 0 \quad \text{for } z < x_{(1)},$$
$$= \frac{j}{n} \quad \text{for } x_{(j)} \leq z < x_{(j+1)} \quad (j = 1, 2, \ldots, n-1),$$
$$= 1 \quad \text{for } x_{(n)} \leq z.$$

If the distribution function of the distribution from which our sample was drawn is really F, then this empirical distribution function F_n should be an approximation to F. Hence it seems sensible to base a critical region of a test of F on the distance of F_n from F. This of course leaves the problem of choosing a measure of the distance between two functions. The Kolmogorov–Smirnov test uses the statistic

$$D_n(x) = \sup_z |F_n(z) - F(z)|,$$

and has critical region of the form

$$\{x : D_n(x) > k\},$$

the constant k being chosen to give the desired size of test. Thus for a size-α test, k is chosen so that

$$\Pr\{D_n(x) > k | F\} = \alpha.$$

The determination of k to satisfy this condition is a problem in probability calculus. Apparently we have a new probability problem for every new F which we wish to test, and a non-trivial problem at that. However it transpires that it is necessary to solve this problem only for one F, since as we vary F, the number k such that $\Pr\{D_n(x) > k | F\} = \alpha$ remains fixed. The statistic $D_n(x)$ is said to be *distribution-free*. We shall not go into the details of a proof of this result. The reader may find it an interesting exercise in probability calculus, given the hint that the critical result is this: if u is a real random variable with continuous distribution function F, then $F(u)$ is uniformly distributed on the interval $[0, 1]$.

The distribution of $D_n(x)$ has been calculated and tabulated for small values of n, so that this test is quite easy to apply.

Note that, in Neyman–Pearson terms, the alternative to the null-hypothesis here consists of a very large set of possible distributions. At least in theory we may calculate the power of the test with respect to any specific alternative. It is also plausible that as $n \to \infty$ the power with respect to any specific alternative tends to 1, so that the test is a good 'all-purpose' test. However greater power may be achieved against particular smaller sets of alternatives by different tests. We have not used any criteria of optimum behaviour in deriving this test.

9.2 The χ^2 goodness-of-fit test

An alternative approach to the problem of the previous section is to reduce the problem to a parametric one by throwing away some of the information contained in the observation x. The line is divided into class intervals by numbers $a_1 < a_2 < \ldots < a_s$; so that we have $s+1$ intervals $I_0 = (-\infty, a_1]$, $I_1 = (a_1, a_2]$, $I_{s-1} = (a_{s-1}, a_s]$, $I_s = (a_s, \infty)$.

The number n_i of sample values falling in the ith class interval is recorded, $i = 0, 1, 2, \ldots, s$, and the original x_is ignored. In this sense we throw away information.

Now whatever the true distribution from which the original sample was drawn, the distribution on the space of $(n_0, n_1, \ldots, n_s)$ is multinomial. If we know nothing at all about the original distribution, the family of possible distributions on $(n_0, n_1, \ldots, n_s)$-space is the family of multinomial distributions which can be labelled by the parameter $\theta = (\theta_0, \theta_1, \ldots, \theta_s)$, where $0 \leq \theta_i \leq 1$ and $\sum_{i=0}^{s} \theta_i = 1$. If the original distribution has distribution function F, then θ has the specific value θ^* say,

where $\quad \theta_i^* = F(a_{i+1}) - F(a_i) \quad (i = 1, 2, \ldots, s-1)$,
$\quad \theta_0^* = F(a_1)$
and $\quad \theta_s^* = 1 - F(a_s)$.

The original problem is in this way reduced to the parametric problem of testing whether the multinomial parameter θ takes the value θ^*. We have already seen how the χ^2 test may be used in such a problem, *if n is large*, and we reject the hypothesis $\{\theta^*\}$ if

$$\sum_{i=0}^{s} \frac{(n_i - n\theta_i^*)^2}{n\theta_i^*}$$

is greater than or equal to the upper 100α per cent point of a χ^2-distribution with s degrees of freedom. Rejection of the hypothesis $\{\theta^*\}$ implies rejection of the original hypothesis $\{F\}$.

The fact that we have thrown away information in order to reduce the problem to parametric form is reflected in the fact that the χ^2-test will have small power against certain of the possible alternatives to $\{F\}$. It is clear that what this approach is really doing is to partition the original family of possible distributions on the sample space into equivalence classes ($F \sim G$ iff $F(a_i) = G(a_i)$, $i = 1, 2, \ldots, s$) and then to provide a test for one of these equivalence classes.

9.3 The Wilcoxon test

The problem is one referred to on page 139 namely, given independent random samples $x = (x_1, x_2, \ldots, x_m)$ and $y = (y_1, y_2, \ldots, y_n)$ from two continuous distributions which are assumed to be the same apart from a possible difference in location, to test whether they do differ in location.

In our general notation, let $\mathscr{F}$ denote the class of continuous distribution functions on the line. Then we may label the family of possible distributions on the sample space by a parameter $\theta = (F, \mu)$, and $\Theta = \{\theta : F \in \mathscr{F}, -\infty < \mu < \infty\}$.

F is the distribution function of each x_i; F^*, where $F^*(\lambda) = F(\lambda - \mu)$, is the distribution function of each y_i. We wish to test the hypothesis ω, where $\omega = \{\theta : F \in \mathscr{F}, \mu = 0\}$.

The idea underlying the Wilcoxon test is that if we order the $m+n$ values observed, there will be no tendency for a preponderance of ys at either end of the ordering if ω is true, but there will be some such tendency if $\mu \neq 0$, that is, if $\Theta - \omega$ is true. In fact, let $z_1, z_2, \ldots, z_{m+n}$ be the ordered set of observed values so that each z is either an x_i or a y_j. Let $r_1, r_2, \ldots, r_n$ be the *ranks* of y_js in the ordered set; that is $r_1, r_2, \ldots, r_n$ are those values of r for which z_r is a member of $y_1, y_2, \ldots, y_n$. Then the Wilcoxon test rejects ω if the statistic

$$R = r_1 + r_2 + \ldots + r_n$$

takes big enough or small enough values, that is, if R lies outside an interval $[k_1, k_2]$.

Again the choice of k_1 and k_2 to ensure a size-α test poses a problem in probability calculus, but not a severe one in this instance. For, when ω is true (whatever the true F), all sets of n ranks $(r_1, r_2, \ldots, r_n)$ are equally probable There are $\binom{m+n}{n}$ such sets and so the probability of each is $\binom{m+n}{n}^{-1}$, under the null-hypothesis. Therefore if k of the $\binom{m+n}{n}$ possible sets of ranks of the y_js are taken as critical region of a test, the test is similar of size $\alpha' = k\binom{m+n}{n}^{-1}$. For a Wilcoxon test of significance level α, we assign to the critical region first those rankings for which R takes its maximum and minimum values, then those for which it takes its next-to-maximum and next-to-minimum values and so on until the number of pairs of rankings assigned, k', say, is such that $2k'\binom{m+n}{n}^{-1}$ is as near α as possible without exceeding it.

Various results are available regarding optimal properties of this test. If the principle of invariance is invoked it can be shown that the original observation (x, y) can be reduced to the set $(r_1, r_2, \ldots, r_n)$ of ranks of the y_js without losing any information which is relevant to the problem in hand. Of course there does not exist a U.M.P. test or even a U.M.P. invariant test of ω. However the reader may refer to Lehmann (1959), p. 236, for further results on the nature of alternatives with respect to which this kind of test is 'good', and for comparisons of the power of this test with the power of its obvious parametric competitor, namely the t-test, which is based on the assumption of normality. Despite the fact that the Wilcoxon test uses only the ranks of the observed values, it fares remarkably well in such comparisons.

9.4 Permutation tests

A commonly occurring problem is this: We are given a set of N observations $(x_1, x_2, \ldots, x_N)$. Under some null hypothesis to be tested the conditional distribution over the set of $N!$ points obtained by permuting the coordinates $x_1, x_2, \ldots, x_N$ assigns equal probability to each of these points, whereas under some alternative of interest this is not so. We may then construct a test of the null hypothesis by limiting consideration to the subset of the sample space consisting of these $N!$ points and choosing a critical region within this set. If there are k points in the critical region the size of the test is $k/N!$. How we assign points to the critical region depends on the alternatives to the null-hypothesis for which we wish reasonable power.

We may illustrate this technique by considering the problem of the previous section where we were given $m+n$ observed values denoted by $x_1, x_2, \ldots, x_m$, $y_1, y_2, \ldots, y_n$. Let $u = (u_1, u_2, \ldots, u_{m+n})$ be a vector obtained by permuting the $m+n$ observed values. Under the null-hypothesis that the xs and ys constitute a random sample from the same distribution, all permutations u are equally likely. Under the alternative that the last n observations constitute a random sample from a distribution shifted relative to that from which the first m arose, then, if the shift is to the right, permutations for which $\bar{u}_n - \bar{u}_m > 0$ are more probable than those for which $\bar{u}_n - \bar{u}_m \leqslant 0$,

$$\text{where} \quad \bar{u}_m = \frac{1}{m} \sum_{i=1}^{m} u_i \quad \text{and} \quad \bar{u}_n = \frac{1}{n} \sum_{i=m+1}^{m+n} u_i;$$

while if the shift is to the left the reverse is true. If we wish a test to guard against both alternatives – a shift to the right and a shift to the left – then it is natural to assign to the critical region, these permutations u for which $|\bar{u}_n - \bar{u}_m|$ is large, the critical value being chosen to yield a test whose size is as near as possible to a predetermined significance level. If we wish to guard only against a shift to the right, that is, if we want reasonable power for this kind of alternative only, then we should assign to the critical region those permutations u for which $\bar{u}_n - \bar{u}_m$ is large. For small samples this procedure may be carried out by enumerating the permutations and calculating $\bar{u}_n - \bar{u}_m$ for each. For large samples it can be shown that under the null hypothesis, the distribution over permutations of $\bar{u}_n - \bar{u}_m$ is approximately normal with zero mean and variance

$$s^2\left(\frac{1}{m} + \frac{1}{n}\right), \quad \text{where} \quad s^2 = \frac{1}{m+n}\left[\sum_{i=1}^{m}(u_i - \bar{u}_m)^2 + \sum_{i=m+1}^{m+n}(u_i - \bar{u}_n)^2\right],$$

provided that the observed values are not dominated by a few of them; this enables a critical region of given size to be determined without vast enumeration.

9.5 The use of a sufficient statistic for test construction

The use of permutations in the way outlined above illustrates a general method of test construction, and, since it is possibly helpful to view permutation tests in this general light, we shall now discuss it. The reader should bear in mind, however, that while we discuss this method in a non-parametric context, it is useful for certain parametric problems also.

The setting is as usual: a sample space X and a family $\{P_\theta : \theta \in \Theta\}$ of distributions on X. We wish to test a null-hypothesis ω against the alternative $\Theta - \omega$.

In section 2.3.2, we defined sufficiency of a statistic t for a family $\{P_\theta : \theta \in \Theta\}$ of distributions. Now suppose that there exists a statistic t which is sufficient for the family $\{P_\theta : \theta \in \omega\}$ but *not* sufficient for the larger family $\{P_\theta : \theta \in \Theta\}$. In other words, the conditional distribution $P_{\theta|t}$ on the sample space is the same for all θ in ω, but not the same for all θ in Θ. If, in this situation, we limit consideration initially to observations x for which $t(x)$ is a given constant, t_0 say, and think in terms of the family $\{P_{\theta|t_0} : \theta \in \Theta\}$ of conditional distributions on the sample space, then the problem of constructing a size-α test of ω against $\Theta - \omega$ is simplified, because relative to this family of conditional distributions ω is a simple hypothesis. It is easier to construct tests of a given size for simple than for composite hypotheses.

If $R(t_0)$ is a region of that subset of X for which $t(x) = t_0$, having the property that

$$P_{\theta|t_0}[R(t_0)] = \alpha, \quad \text{for } \theta \in \omega,$$

we say that $R(t_0)$ is the critical region of a size-α *conditional* test of ω against $\Theta - \omega$.

If, for each possible value of t_0, we have constructed such a critical region $R(t_0)$ (which in practice may be quite easy) and if R is the union of the $R(t_0)$ over all possible t_0, then R is the critical region of a *similar* size-α test of ω

against $\Theta - \omega$,

since
$$\begin{aligned}P_\theta(R) &= E_\theta\{P_{\theta|t}(R)\} \\ &= E_\theta\{P_{\theta|t}[R(t)]\} \\ &= E_\theta(\alpha) \quad \text{for all } \theta \in \omega \\ &= \alpha.\end{aligned}$$

As an illustration of this technique for constructing similar tests via conditional tests, let us consider the following problem. Given the results of n independent trials we wish to test the null-hypothesis that the probability of success remains constant over the n trials against the alternative that this probability increases at least once during the course of the trials.

Here our sample space X is the set of 2^n elements $x = (x_1, x_2, \ldots, x_n)$, where each x_i is either 1 (for success) or 0 (for failure). The parameter space Θ is a subset of R^n,

$$\Theta = \{\theta = (\theta_1, \theta_2, \ldots, \theta_n) : 0 \leq \theta_i \leq 1, \theta_1 \leq \theta_2 \leq \ldots \leq \theta_n\},$$

and the null hypothesis ω is given by

$$\omega = \{\theta \in \Theta : \theta_1 = \theta_2 = \ldots = \theta_n\}.$$

As we have seen in chapter 2, the statistic

$$t(x) = \sum_{i=1}^{n} x_i$$

is sufficient for the family $\{P_\theta : \theta \in \omega\}$ on X, where generally

$$P_\theta(x) = \prod_{i=1}^{n} \theta_i^{x_i}(1-\theta_i)^{1-x_i}$$

and, when $\theta = (\gamma, \gamma, \ldots, \gamma)$, so that $\theta \in \omega$,

$$P_\theta(x) = \gamma^{\Sigma x_i}(1-\gamma)^{n-\Sigma x_i}.$$

However, it is intuitively clear that the order of occurrence of the successes (as opposed to their total number) gives some information about whether the probability of success increases during the course of the trials, and indeed t is *not* sufficient for the family $\{P_\theta : \theta \in \Theta\}$ (see section 2.3.1).

Corresponding to a given value of t, $t(x) = r$, say, there are $\binom{n}{r}$ points in the sample space and, for each $\theta \in \omega$, the conditional θ-distribution over these points assigns equal probability to each, as is readily verified. For any $\theta \notin \omega$, the conditional θ-distribution does not assign equal probability to each of those $\binom{n}{r}$ points. More probability is assigned for instance to the x for which

$$x_1 = x_2 = \ldots = x_{n-r} = 0 \quad \text{and} \quad x_{n-r+1} = x_{n-r+2} = \ldots = x_n = 1$$

than to that for which

$$x_1 = x_2 = \ldots = x_r = 1 \quad \text{and} \quad x_{r+1} = x_{r+2} = \ldots = x_n = 0.$$

Hence for a size-α conditional test of ω it is natural to assign to the critical region, those of the $\binom{n}{r}$ possible xs for which the sum of the ranks of the components which are 1 is larger than a chosen constant, depending on α. (Of course, because of the discrete nature of the problem, a randomized test might be necessary to achieve the exact size α; but this is irrelevant to the main idea.)

For each given r, that is, each possible value of the statistic $t(x)$, we may construct a size-α conditional test in this way and so build up a critical region in the whole sample space X. According to the above general argument, this is the critical region of a similar size-α test of ω against $\Theta - \omega$.

A question of considerable interest is this: suppose that each size-α conditional test is optimal in some sense – is, for instance, U.M.P. in the class of size-α conditional tests. Does it follow that the similar size-α test built up from these conditional tests has optimal properties? We shall not pursue this question here. Once again the reader is referred to Lehmann (1959), section 4.3, for a discussion of it. Instead we shall focus attention on how permutation tests fit into this general framework.

Let $X = R^N$, so that a typical observation is $x = (x_1, x_2, \ldots, x_N)$. We consider a family of distributions on R^N, each of which possesses a density function with respect to Lebesgue measure. We may then label this family by their density functions, so that if, as usual, we use θ as a label, θ is a density function, $\theta(x_1, x_2, \ldots, x_N)$, and Θ is a family of density functions. Let $\mathscr{F}$ be the family of probability density functions on the line and let ω be the family of distributions on R^N labelled by parameters θ of the form

$$\theta(x_1, x_2, \ldots, x_N) = \prod_{i=1}^{n} f(x_i), \quad f \in \mathscr{F}.$$

If we think of ω as a hypothesis it is that which says that $x_1, x_2, \ldots, x_N$ are continuous independent identically distributed random variables.

Suppose that Θ contains ω and that we wish to test the null hypothesis ω against $\Theta - \omega$.

Let $x^{(1)} \leqslant x^{(2)} \leqslant \ldots \leqslant x^{(N)}$ denote the ordered values of $x_1, x_2, \ldots, x_N$ and consider the statistic t defined by

$$t(x_1, x_2, \ldots, x_N) = (x^{(1)}, x^{(2)}, \ldots, x^{(N)}),$$

the so-called *order statistic*. Then by the factorization theorem of chapter 2, t is sufficient for the family $(P_\theta : \theta \in \omega)$ of distributions on the sample space R^N.

If we are given a value of t, say,

$$t(x) = (a_1, a_2, \ldots, a_N),$$

where $a_1 < a_2 < \ldots < a_N$, then there are $N!$ points in the sample space corresponding to this value of t, namely those whose coordinates are permutations of $a_1, a_2, \ldots, a_N$. For each $\theta \in \omega$, the conditional θ-distribution over these $N!$ points assigns equal probability to each. (We may ignore those values of t which contain at least two equal components, since for every $\theta \in \Theta$, the probability of the set of such values of t is zero.)

Using given values of t we may now construct size-α conditional tests of ω just as in the above illustration, and this is how permutation tests fit into the general framework under discussion.

9.6 Randomization

9.6.1 Permutation tests underline the principle of randomization in experimentation whereby experimental units are assigned to various treatments by some random mechanism, and we illustrate this point by considering two examples.

Suppose that it is desired to test whether r different fertilizers are equally effective and rs plots are available for experimentation. To help avoid the possibility that some treatments are assigned to specially favourable plots, that is, to plots with natural high fertility, each treatment is applied to s randomly chosen plots. Let the results be as follows:

Treatment	Yield			
1	x_{11}	x_{12}	$\ldots$	x_{1s}
2	x_{21}	x_{22}	$\ldots$	x_{2s}
$\vdots$	$\vdots$	$\vdots$		$\vdots$
r	x_{r1}	x_{r2}	$\ldots$	x_{rs},

and let $z = (z_{ij}; i = 1, 2, \ldots, r, j = 1, 2, \ldots, s)$ be a permutation of the observed values $(x_{ij}; i = 1, 2, \ldots, r, j = 1, 2, \ldots, s)$. Under the null-hypothesis of no differences among treatments each permutation z has probability $1/(rs)!$. Whereas under the alternative that the treatments simply shift a basic distribution, those permutations z for which

$$f(z) = \frac{\sum_i (z_{i.} - z_{..})^2}{\sum_{i,j} (z_{ij} - z_{i.})^2}$$

is large are more probable. Here $sz_{i.} = \sum_j z_{ij}$ and $rsz_{..} = \sum_i \sum_j z_{ij}$. Hence we obtain a test of the null-hypothesis of size

$$\alpha = \frac{k}{(rs)!}$$

by assigning to the critical region those k permutations z for which $f(z)$ takes its k largest values; and this test guards particularly against alternatives which state that treatment effects can be described by 'shifts' in a basic distribution.

A parametric approach to this problem might set up the model

$$x_{ij} = \mu_i + \varepsilon_{ij}, \quad (i = 1, 2, \ldots, r, \quad j = 1, 2, \ldots, s)$$

where the μs are unknown constants and the εs are independent random variables each of which is assumed to have an $N(0, \sigma^2)$ distribution, with σ^2 unknown. Then the null hypothesis specifies that $\mu_1 = \mu_2 = \ldots = \mu_r$, and the likelihood-ratio test can be applied. The critical region of this test turns out to be of the form $\{x = (x_{ij}) : f(x) > c\}$, as the reader may verify, where f is the statistic on which we based the permutation test. This is no accident. Rather it illustrates a method of determining a critical region for a permutation test in order to guard against certain kinds of alternative. We set up a parametric model, usually based on the assumption of normality, determine a critical region for the parametric test which is optimum in some sense and then use the same statistic to construct the critical region of a permutation test. The parametric approach gives us a guide concerning the 'shape' of the critical region which we should adopt to achieve power where we want to achieve it. And it often does more. For it often happens that for large samples (large s in the above example) the permutation distribution of the statistic used is, subject to weak assumptions, approximately the same as its distribution derived under the assumption of normality. This is true for instance in the above example, and if s is reasonably large, we need not become involved in large scale enumeration in order to determine the permutations which comprise the critical region of a test of given size.

Another way of designing the experiment concerning r fertilizers and rs plots is to divide the rs plots into s *blocks* each of r plots, the plots in any one block being chosen to be as homogeneous as possible. Then treatments are randomized within each block, that is the r plots in each block are assigned to the r treatments at random. Suppose that the results are now:

	Block			
	1	2	...	s
Treatment				
1	x_{11}	x_{12}	...	x_{1s}
2	x_{21}	x_{22}	...	x_{2s}
⋮	⋮	⋮		⋮
r	x_{r1}	x_{r2}	...	x_{rs}

Let $u = (u_{ij}; i = 1, 2, \ldots, r; j = 1, 2, \ldots, s)$ be a set of numbers obtained from the above x_{ij}s by permuting within columns, so that, for each fixed j, $(u_{1j}, u_{2j}, \ldots, u_{rj})$ is a permutation of $(x_{1j}, x_{2j}, \ldots, x_{rj})$. There are $s(r!)$ such us, and because of the method of randomization employed they are equally likely under the null hypothesis of no differences among treatments. We therefore obtain a test of this hypothesis of size

$$\alpha = \frac{k}{s(r!)}$$

by assigning k of the us to a critical region. The criterion for choosing the us to be assigned to the critical region of a test depends on the alternatives to the null-hypothesis which we wish particularly to guard against. And again a parametric model may be used as a guide, both to suggest the statistics to be used and (possibly) to yield a large-sample (large s) approximation to its permutation distribution. Suppose, for instance, that we wish reasonable power for alternatives which specify that treatments differ only in shifting the s basic distributions (one for each block). Then we might consider the parametric model

$$x_{ij} = \mu + \beta_j + \tau_i + \varepsilon_{ij},$$

where μ, β_j ($j = 1, 2, \ldots, s$) and τ_i ($i = 1, 2, \ldots, r$) are unknown parameters and ε_{ij} ($i = 1, 2, \ldots, r$; $j = 1, 2, \ldots, s$) are independent $N(0, \sigma^2)$ random variables. Here the βs are 'block constants' and the τs 'treatment constants'; for identification we impose the conditions

$$\sum \beta_j = 0, \qquad \sum \tau_i = 0.$$

We now consider testing the null-hypothesis that the τs are all equal against the alternative that they are not – this corresponds to testing the hypotheses that the treatments are equally effective against the alternative that they differ by shifting distributions. Application of the likelihood-ratio test (which can be shown to have optimum properties for this particular problem) produces as test statistic

$$f(x) = \frac{\sum_i (x_{i.} - x_{..})^2}{\sum_{i,j}(x_{ij} - x_{..})^2 - r\sum_j (x_{.j} - x_{..})^2 - s\sum_i (x_{i.} - x_{..})^2},$$

where the dot denotes average over the index which it replaces. Furthermore, under the null hypotheses $r(s-1)f(x)$ has an F-distribution with $(r-1)$ and $(r-1)(s-1)$ degrees of freedom.

We may use this statistic for a permutation test also, and put into the critical region of such a test, those us for which $f(u)$ is large. If s is large we may use the F-distribution quoted as an approximation to the *permutation* distribution of f and thereby avoid enumeration.

9.6.2. We have considered here only a few examples of the many non-parametric tests in existence. For others the reader is referred to Walsh (1962).

Moreover we have not discussed any theory of optimal properties of non-parametric procedures. This has been developing in recent years, and the reader is referred to Lehmann (1959), Hajek and Sidek (1967), and Noether (1967) for discussion of this type of problem and for further references.

Examples

9.1 The following observations are a random sample from some continuous distribution:

3·02	6·12	4·26	4·24
9·42	5·56	2·18	7·20
0·36	6·82	3·80	1·18
3·90	2·46	8·50	4·72
6·20	2·86	3·74	2·30

Sketch the empirical distribution function. On the same diagram sketch the distribution function of a normal distribution with mean 5 and standard deviation 2. May it be assumed that the sample was drawn from this distribution? (See Lindgren, 1962, for tabulated critical values of the Kolmogorov test.)

9.2 An investigation was carried out on two suggested antidotes to the consequences of drinking, these being (a) 2 lb of mashed potatoes and (b) a pint of milk. Ten volunteers were used, five to each antidote, the allocation to antidote being random. One hour after each had drunk the same quantity of alcohol and swallowed the appropriate antidote, a blood test was carried out and the following levels (mg/ml) of alcohol in the blood were recorded:

(a) 76 52 92 80 70
(b) 110 96 74 105 125.

By means of a non-parametric test, decide whether there is sufficient evidence to conclude that one treatment is more effective than the other.

9.3 A sample of maggots is placed on an enclosed plate. These wander about for a time and eventually each becomes static and turns into a chrysalis. When this stage has been reached, the position of each chrysalis on the plate is noted. There are three main possibilities: (a) the chrysalides are distributed randomly on the plate; (b) there is a tendency for them to cluster; (c) they tend to isolate themselves one from another. How would you assess the evidence provided by the data collected regarding these three possibilities?

9.4 The following table gives efficiency indices for each of two large firms and two small firms chosen at random from each of three areas.

Area	Large	Small
1	22, 20	16, 9
2	19, 11	12, 15
3	11, 8	6, 4.

Using non-parametric methods, discuss the evidence provided by these data regarding the question of differences in efficiency between large and small firms.

9.5 Let $x_1, x_2, \ldots, x_{2n}$ be a random sample from a distribution with unknown mean μ and consider the following test of the hypothesis H that $\mu = 0$. Reject H if $|r-n| > k$, where r is the number of positive observations.

Assuming that the distribution is symmetric, determine k so that this test has approximate size α.

Calculate the power of the test for the alternative that the distribution is $N(\mu, \sigma^2)$ and compare this power with that of the size-α t-test having critical region

$$|\bar{x}| > c\sqrt{\{\sum(x_i - \bar{x})^2\}}.$$

9.6 The following table gives the results of a paired comparison experiment between two treatments. (Experimental units in a pair are chosen to be as similar as possible to one another.)

Treatment	Pair 1	2	3	4	5	6	7	8	9	10
1	24	18	30	25	16	17	21	18	20	14
2	22	15	26	24	32	16	19	17	18	13
Differences	+2	+3	+4	1	−16	+1	+2	+1	+2	+1

Carry out (a) a t-test, (b) a sign test, each of size 0·05, on the differences to test the hypothesis that there is no difference between treatments. These tests result in different conclusions. Which one do you believe and why? What may this example illustrate regarding the powers of the two tests?

10 The Bayesian Approach

In previous chapters we have attempted to develop a theory of inference for certain kinds of problem in terms of probabilities which admit a 'frequency' interpretation; that is, we have associated probabilities only with events arising from experiments which can be repeated. Then a probability can be interpreted roughly as the relative frequency of occurrence of the event in a large number of repetitions of the appropriate experiment. We have *not* used probabilities to describe degrees of belief in possible alternative parameter values (or states of nature) and we have never made statements such as 'the probability that this hypothesis is true is so and so', because the truth of a hypothesis is not an event arising from a repeatable experiment.

Referring to the problem of set estimation in the introductory chapter we said that this was the problem of dividing the set of possible parameters into two subsets, one of which was 'plausible' *ex post* (that is, after an observation was made) and one implausible *ex post*. We also said that we would be concerned with clarifying the sense in which these subsets were respectively plausible and implausible. Some statisticians argue that the explanation of a confidence set provided by the 'frequentist' approach, far from clarifying the sense in which a subset is plausible, merely involves mental juggling which obscures the whole issue; that degrees of belief are involved in a much more detailed way than we have admitted in confidence statements with a frequency interpretation; and that by accepting this fact we arrive at a much more realistic analysis of the process of scientific inference. We leave the reader to consider this argument for himself with the comment that the mere use of the word plausible in our introduction lends some support to it.

10.1 Prior distributions

The essential difference between the approach to inference which we have adopted until now and the alternative approach, called the Bayesian approach which does not shrink from using probabilities to specify degrees of belief, is that in the latter there is one further ingredient in a general mathematical model. As before, we have a sample space X, a parameter space Θ and a family $\{P_\theta : \theta \in \Theta\}$ of probability distributions on X – a family of possible states of nature as it is sometimes described. However now we assume the existence of a probability distribution Π on a class of measurable sets *in* Θ. This probability distribution describes our degrees of belief in possible parameter values prior to an observation being made, and consequently it is called a *prior distribution*. We defer for the moment any discussion of how to arrive at the appropriate prior distribution for any given problem. Even if a purely pragmatic attitude is adopted it does seem to be true that for at least some inference problems, an approach which assumes the existence of a prior distribution is more realistic than one which does not. Let us assume, then, that we have this further ingredient in our mathematical model. How does the theory of inference proceed?

10.2 Posterior distributions

The fact that the probability distribution Π has a different interpretation in practice from other probability distributions which we have introduced does not make it a different *mathematical* entity. We can conjoin it with other distributions exactly as if it had the same interpretation. In particular we may establish a 'joint distribution of x and θ', that is, a distribution on $X \times \Theta$, by thinking of P_θ as a conditional distribution on X given θ. To emphasize this it is convenient to change notation once again and write $P(\cdot|\theta)$ instead of P_θ. Now suppose that, for each θ, $P(\cdot|\theta)$ is defined by a density function $p(\cdot|\theta)$ with respect to some fixed measure on X and that Π is defined by a density function π with respect to some fixed measure on Θ. Then the density function $p(x, \theta)$, with respect to the product of the fixed measures, of the joint distribution of x and θ is given by

$$p(x, \theta) = \pi(\theta) p(x|\theta).$$

(Note that here we are using p as a generic notation for density function and not for a fixed function.)

Usually we can derive from this joint distribution, a conditional distribution of θ given x, defined by the density

$$p(\theta|x) = \pi(\theta) \frac{p(x|\theta)}{p(x)},$$

where $\quad p(x) = \int_\Theta \pi(\theta) p(x|\theta) \, d\theta,$

and is the density of the marginal distribution of x. This conditional distribution on Θ may be interpreted as describing our degrees of belief in different possible values of θ after the observation x has been made, and consequently it is called the *posterior distribution* of θ.

Once a prior distribution is given, the whole inference process is very easily summed up according to this approach: *An observed result changes our degrees of belief in different parameter values by changing a prior distribution into a posterior distribution.*

In a sense this says all that need be said about inference, if we accept the existence of a prior distribution and the above very natural interpretation of a posterior distribution. It may be that in practice we wish to concentrate attention on some particular aspect of the posterior distribution, so that it is not necessary to communicate it in its entirety. However it is usually clear from the context how we use the posterior distribution to provide the summary required.

10.3 Bayesian confidence intervals

10.3.1 To illustrate the point just made, suppose that we wish to provide a confidence set for an unknown real parameter $g(\theta)$. From the posterior distribution of θ we may calculate the posterior distribution of g. Suppose, for the sake of illustration, that this has density $p(g|x)$ and that the graph of $p(\cdot|x)$ is as shown in Figure 6.

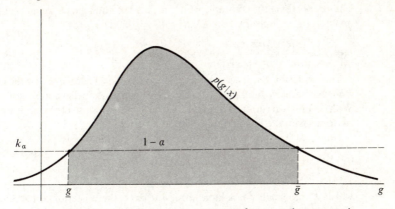

Figure 6 Bayesian confidence interval $(\underline{g}, \bar{g})$ for an unknown real parameter g

The obvious way to construct a $100(1-\alpha)$ per cent *Bayesian* confidence interval, that is, an interval for g whose posterior probability is $1-\alpha$, is to find the constant k_α such that

$$\int_{\{g:p(g|x) > k_\alpha\}} p(g|x)\, dg = 1-\alpha.$$

Then $\{g:p(g|x) > k_\alpha\}$ is a $100(1-\alpha)$ per cent confidence interval in the sense that our posterior confidence (or degree of belief) that g lies in this interval is $100(1-\alpha)$ per cent. Moreover this interval clearly consists of those values of g most plausible *ex post* – any value outside the interval is less plausible than any value in it. Note that in general we cannot make this latter statement for frequentist confidence intervals, so that if we must use the term plausibility in explanations, the Bayesian analysis must be more satisfactory than the frequentist one.

10.3.2 Example

Suppose that n independent identical trials yield r successes. Our prior degrees of belief in different possible values of θ, the common probability of success, are described by the uniform distribution on $[0, 1]$, that is $\pi(\theta) \equiv 1, 0 \leqslant \theta \leqslant 1$. What is the posterior distribution of θ?

$$P\{r \text{ successes}|\theta\} = \binom{n}{r}\theta^r(1-\theta)^{n-r}.$$

Now the density $p(\theta|r)$ of the conditional distribution of θ given r successes is proportional to $\pi(\theta)p(r|\theta)$ and so

$$p(\theta|r) \propto \theta^r(1-\theta)^{n-r}.$$

The proportionality constant, which makes the total posterior probability unity on $[0, 1]$ is clearly $\{B(r+1, n-r+1)\}^{-1}$ and so

$$p(\theta|r) = \frac{\theta^r(1-\theta)^{n-r}}{B(r+1, n-r+1)}.$$

This posterior density has a maximum at $\theta = r/n$. Therefore r/n is the most plausible value of θ *ex post* – that value in which we believe most strongly. With a uniform prior distribution, the Bayesian most-probable estimate coincides with the maximum likelihood estimate.

Note that the variance of the posterior distribution of θ is

$$\frac{(r+1)(n-r+1)}{(n+2)^2(n+3)},$$

which is very small if n is large. So if n is large, the posterior distribution is highly concentrated whereas the prior distribution which we adopted is very diffuse. This is completely in accord with what one would hope for. Vague prior knowledge is transformed into rather precise posterior knowledge by a very informative experiment.

10.4 Bayesian inference regarding hypotheses

The Neyman–Pearson theory of hypothesis testing is not really concerned with *inference* regarding hypotheses. Rather it seeks to provide a solution of a *decision* problem, a solution which is optimum relative to certain criteria. The Bayesian approach, on the other hand, admits genuine inference for hypotheses.

Suppose that we have two hypotheses ω and $\Theta - \omega$. The prior distribution Π assigns the probability $\Pi(\omega)$ to ω and $1 - \Pi(\omega)$ to $\Theta - \omega$. These probabilities express our prior degrees of belief in the respective hypotheses. After an observation x has been made, the posterior distribution $P(\cdot|x)$ assigns probability $P(\omega|x)$ to ω and $1-P(\omega|x)$ to $\Theta-\omega$, and these express our posterior degrees of belief in the respective hypotheses. This is all that need be said.

Of course we may use the Bayesian approach to decision problems also; the Bayesian attitude here is that the posterior distribution $P(\cdot|x)$ specifies our state of knowledge about θ after the observation x which is to help in decision making, and that such a specification is a necessary preliminary to the solution of any decision problem involving θ. We shall consider the role in decision problems of the posterior distribution in the next chapter.

10.5 Choosing a prior distribution

The reader may find the Bayesian theory of inference intellectually more satisfying than that provided by the frequentist approach which we have considered until now. The latter approach leaves the suspicion that its justifications of intuitively acceptable solutions are very much *ad hoc* and that it involves changing ground slightly from one such solution to the next. On the other hand, once we have accepted that degrees of belief can properly be described by probability distributions and *we have established a method of determining the appropriate prior distribution for each problem we encounter*, the Bayesian approach provides a general method for solving all problems of inference. The rub lies in the qualification italicized. As yet we have not mentioned how to arrive at an appropriate prior distribution and this is the subject of some controversy.

It is sometimes argued that by sufficient introspection one can arrive at the prior odds at which one would just accept a bet on this parameter value rather than that, and so eventually find *the* prior distribution appropriate for a particular problem. This may well be true in theory but of course in practice it would result in a life of introspection in which there was no time to make any observation in the real world at all. In fact, in practice, prior knowledge is often rather vague and there is a whole class of prior distributions, each one of which is adequate for describing an individual's degrees of belief. The experimenter can usually distinguish between a prior which is 'good enough' and one which is 'not good enough'. Within the 'good enough' class we can then choose one on, for instance, grounds of mathematical tractability.

This pragmatic attitude is supported by the fact that with a reasonably informative experiment, a prior distribution adequate for describing rather imprecise knowledge can be changed quite considerably without affecting the posterior distribution all that much. Consider, for instance, the case of n independent identical trials discussed in Example 10.3.2. There we adopted a uniform prior distribution for θ and arrived at the posterior distribution with density

$$p(\theta|r) = \frac{\theta^r(1-\theta)^{n-r}}{B(r+1, n-r+1)}.$$

If instead we had used the prior distribution with density

$$\pi(\theta) = 6\theta(1-\theta) \quad (0 \leqslant \theta \leqslant 1),$$

a prior distribution which still describes somewhat vague knowledge of θ but expresses a belief that θ is more likely to be around the middle of the interval $[0, 1]$ than near its extremities, then the posterior distribution would have had density

$$p(\theta|r) = \frac{\theta^{r+1}(1-\theta)^{n-r+1}}{B(r+2, n-r+2)}.$$

When n and r are reasonably large, there is little difference between these posterior distributions. In practice then, with a fairly informative experiment, we need not be over-careful about the choice of prior distribution.

10.6 Improper prior distributions

It is not necessary that a prior distribution be a proper probability distribution in the sense that it assigns probability 1 to the whole parameter space. Suppose for instance that the parameter space is the whole real line and it suits our convenience to think of all parameter values as being 'equally likely' *a priori*. It seems natural to describe this by a 'distribution' with a constant density function

$$\pi(\theta) = k, \quad \text{say, for all } \theta.$$

However for no value of k does this define a proper probability distribution, since $\int_{-\infty}^{\infty} \pi(\theta) \, d\theta$ does not exist. Nevertheless we may use Lebesgue measure and take $\pi(\theta) = 1$, for all θ, to describe our prior degrees of belief. Such a prior distribution is called *improper*. Of course improper prior distributions may, and usually do, result in perfectly proper posterior distributions. Note that in the case where Θ is a Euclidean space, taking an improper prior with constant density is equivalent to using the likelihood function – now written $p(x|\theta)$ – to describe posterior degrees of belief. For example, suppose that $(x_1, x_2, \ldots, x_n)$ is a random sample from an $N(\theta, 1)$ distribution with θ unknown, so that

$$p(x|\theta) = \frac{1}{(2\pi)^{\frac{1}{2}n}} \exp\left\{-\tfrac{1}{2}\sum (x_i - \theta)^2\right\}$$

$$= \frac{\exp\left\{-\tfrac{1}{2}\sum (x_i - \bar{x})^2\right\}}{(2\pi)^{\frac{1}{2}n}} \exp\left\{-\tfrac{1}{2}n(\theta - \bar{x})^2\right\}.$$

With an improper prior distribution for θ, having constant density, we have $p(\theta|x)$ proportional to $p(x|\theta)$ and so

$$p(\theta|x) \propto \exp\left\{-\tfrac{1}{2}n(\theta - \bar{x})^2\right\}.$$

Obviously, then, the posterior distribution of θ is $N(\bar{x}, 1/n)$.

The Bayesian approach may be used to justify most intuitively acceptable *parametric* procedures and for an account of this the reader is referred to Lindley (1965). However, in more complex inference problems there remains the practical difficulty of choosing a prior distribution to reflect one's degrees of belief. This difficulty is highlighted in non-parametric problems, when it is certainly not easy to identify degrees of belief with prior distributions on the function spaces which then constitute the parameter space. The essential simplicity of Bayesian theory is attractive. It is not difficult to apply once a prior distribution has been assigned. But to assign one in practice may be an extremely difficult problem.

In the next chapter we shall see how prior distributions arise naturally in a mathematical context, and this provides considerable support for their use, irrespective of whether or not they are interpreted as describing degrees of belief.

Examples

10.1 Let $x_1, x_2, \ldots, x_n$ be a random sample from an $N(\theta, \sigma^2)$ distribution, where σ^2 is known. Prove that, when the prior distribution of θ is $N(\mu, \tau^2)$, its posterior distribution is $N(\mu_n, \tau_n^2)$,

where $\quad \mu_n = \dfrac{n\bar{x}/\sigma^2 + \mu/\tau^2}{n/\sigma^2 + 1/\tau^2} \quad$ and $\quad \tau_n^{-2} = n\sigma^{-2} + \tau^{-2}$.

10.2 Let $x_1, x_2, \ldots, x_n$ be a random sample from an $N(\theta_1, \theta_2)$ distribution, with both θ_1 and θ_2 unknown, and suppose that the prior distribution of $\theta = (\theta_1, \theta_2)$ has 'density'

$$\pi(\theta) \propto 1/\theta_2 \quad (-\infty < \theta_1 < \infty, \theta_2 > 0).$$

(Thus θ_1 and θ_2 are independent and each has an improper prior distribution, θ_1 being uniform on $(-\infty, \infty)$ and θ_2 having 'density' $1/\theta_2$ on $(0, \infty)$.) Show that the posterior distribution of θ_1 is such that $n^{\frac{1}{2}}(\theta_1 - \bar{x})/s$ has Student's t-distribution with $n-1$ degrees of freedom,

where $\quad s^2 = \dfrac{1}{n-1} \sum_{i=1}^{n} (x_i - \bar{x})^2$.

Then find a shortest 95 per cent Bayesian confidence interval for θ_1.

10.3 A physical system has two possible states θ_1 and θ_2 and its progression from one state to another is described by a homogeneous Markov chain with transition matrix

$$P = \begin{bmatrix} \alpha & 1-\alpha \\ 1-\beta & \beta \end{bmatrix},$$

where $0 < \alpha, \beta < 1$. If the prior probability is π_1 that the initial state is θ_1, determine the posterior probability that the initial state was θ_1, given that the system occupies state θ_1 after n steps. Explain why this posterior probability approaches π_1 as n becomes large.

10.4 From a production run a large number of electric light bulbs is obtained, the variability in their lifetimes being described by a distribution with density $\theta e^{-\theta x}(x > 0)$. A run is regarded as satisfactory if the mean lifetime is greater than t. The parameter θ varies from batch to batch depending on the quality of the tungsten used for the filaments and this variability in θ is described by a distribution with density $[\Gamma(k)]^{-1}\theta^{k-1}e^{-\theta}(\theta > 0)$. From a production run, n bulbs chosen at random have lifetimes $x_1, x_2, \ldots, x_n$. On the basis of this information, find the probability that the run has been satisfactory.

10.5 An event E has an unknown probability θ, believed to be small, of occurring at each of a sequence of independent trials. To obtain information about θ, trials are carried out until E occurs for the first time, which happens at the nth trial. If a prior distribution with density $m(1-\theta)^{m-1}(0 < \theta < 1)$ is assigned to θ, show that the shortest Bayesian confidence interval for θ having probability $1-\alpha$ is, for large $m+n$, given to a good approximation by

$$\left[\frac{c_1}{(m+n)}, \frac{c_2}{(m+n)}\right],$$

where $c_1 e^{-c_1} = c_2 e^{-c_2}$, $e^{-c_1} - e^{-c_2} = 1-\alpha$. (*Camb. Dip.*)

10.6 Given a uniform prior distribution for the probability θ of success in each of a sequence of independent trials, show that the probability that n trials yield r successes is $\dfrac{1}{n+1}$, $r = 0, 1, \ldots, n$.

If it is known that the first n trials have produced r successes, show that the probability is $(r+1)/(n+2)$ that the next trial will result in success.

10.7 The number of particles emitted in time T from a radioactive source may be assumed to have a Poisson distribution with mean λT, where λ is the emission rate of the source. Counts x and y are made in time T from two independent sources whose rates λ_1 and λ_2 are unknown. Prior to these counts λ_1 and λ_2 are taken to be independently distributed, each with density $e^{-\lambda}(\lambda > 0)$. Compare the prior and posterior probabilities that the ratio λ_1/λ_2 is between $1-\delta$ and $1+\delta$, when x and y are both large, and δ is small.

11 An Introduction to Decision Theory

The hypothesis-testing problem is really a decision problem rather than an inference problem. As a result of an observation which is informative in some sense, we have to reach one of two alternative decisions. The Neyman–Pearson theory provides an analysis of this problem which is to some extent half hearted. It recognizes that the two possible errors involved in a decision procedure, that is, in a test, are not equally serious and the criteria that it lays down for a 'good' test are based on this notion. But it does not involve any detailed analysis of the 'seriousness' of different errors. If we do carry out such an analysis, we are led naturally to the following interpretation of the hypothesis-testing or two-decision problem.

11.1 The two-decision problem

11.1.1 After making an observation we have to make one of two decisions:

d_0: a hypothesis ω regarding an unknown parameter θ is true;

d_1: the hypothesis ω is false.

We may analyse the consequence of the decision d_i when θ is the true state of nature and in this way arrive at a loss function (a gain being a negative loss), $L(d_i, \theta)$, which expresses the seriousness of the errors which we may make. This loss function constitutes a new ingredient in our mathematical framework.

Clearly we shall wish to minimize loss in some way, and our problem becomes this: to choose a decision procedure, that is, a rule which assigns one of the two decisions d_0 and d_1 to each point of the sample space, and to choose this so that loss is minimized in a way to be specified. From a mathematical point of view a decision procedure is simply a function from the sample space into the 'decision space' of two elements d_0 and d_1. Thus our problem becomes that of choosing a decision function.

11.1.2 One of the virtues of this approach to the hypothesis-testing problem is that other general problems fit into this same framework. In particular the problem of point estimation does so. To see this, consider the case where we have a single unknown real parameter θ. Then we may regard an estimate $\hat{\theta}(x)$ as a decision, and an estimator $\hat{\theta}$ as a decision function. Again we may analyse the consequences of 'wrong decisions' and arrive at a loss function $L\{\hat{\theta}(x), \theta\}$

which expresses the demerit of the estimate $\hat{\theta}(x)$ when θ is the true state of nature. A typical loss function in this case might well be

$$L\{\hat{\theta}(x), \theta\} = \{\hat{\theta}(x) - \theta\}^2.$$

And again our problem is to choose a decision function (an estimator $\hat{\theta}$) which minimizes loss in some way.

11.2 Decision functions

To investigate general principles and the way in which we wish to minimize loss, we may as well look at the general decision problem of which the problems of estimation and hypothesis testing are two examples. The mathematical ingredients of this general problem are as follows.

As always we have a sample space X whose elements represent possible outcomes of an experiment, and a parameter space Θ whose elements represent possible 'states of nature'. Now, in addition, we have a decision space D, each of whose elements represents a decision which we may make, and a loss function L defined on $D \times \Theta$, $L(d, \theta)$ representing the loss (or gain if $L(d, \theta)$ is negative) resulting from the decision d, when θ is the true state of nature. A decision function δ is a mapping from X into D, and our problem is to choose a δ which is 'good' in some sense: that is, which minimizes loss in some way.

11.3 The risk function

11.3.1 We are now back to this recurrent question, 'What do we mean by good?' As before we can proceed a certain distance in an automatic way, for, given a decision function δ we may calculate the 'risk' $R_\delta(\theta)$ associated with δ when θ is the true state of nature. This risk is defined, for each θ, by

$$R_\delta(\theta) = E_\theta\{L(\delta, \theta)\}.$$

Note that, for each fixed θ, $L(\delta, \theta)$ is a function on the sample space and we may calculate its expected value relative to any distribution on X, in particular, to that determined by θ.

Now for each δ we have a *risk function* R_δ defined on the parameter space, and it may seem natural to define a good decision function as one having uniformly minimum risk. However this is not a very useful definition for practical purposes as it is seldom that such a decision function exists. We have encountered just this difficulty previously in the case of estimation and indeed we may refer to that case to indicate the kind of reason why it is unrealistic in practice to look for a uniformly minimum risk decision function. Suppose again that θ is a real parameter for which we require a point estimate, and that the loss function is quadratic

$$L\{\hat{\theta}(x), \theta\} = \{\hat{\theta}(x) - \theta\}^2.$$

In this case the decision function is an estimator $\hat{\theta}$ and the risk of $\hat{\theta}$ is simply its mean-square error. Thus to demand a decision function of uniformly minimum risk is to demand an estimator with uniformly minimum mean-square error. We have seen in chapter 2 that only in very exceptional circumstances does such an estimator exist, and the reasons given there are just the kind of reasons why in general no uniformly minimum risk decision function exists. We are back at the impasse which we have encountered before.

11.3.2 Example

In order to give us some basis for assessing different suggestions for circumventing this difficulty we consider now a very simple example where it is possible to enumerate all possible decision functions and calculate their risks.

An airline has an option on ten second-hand aircraft, all in similar condition. Of these an unknown number θ will give 1000 hours flying time without major breakdown, and on each of these the airline will make a profit of £1000p: on each of those which do suffer a major breakdown within 1000 hours the airline will lose £1000q. A decision has to be made on whether or not to take up the option and to obtain some information about θ, the airline subjects one of the aircraft to tests from which it emerges satisfactory if it will yield 1000 hours flying time and unsatisfactory if not. (There is time to test only one.) The cost of these tests is £1000r. Determine the risk function of the four possible decision procedures.

For this example the sample space X has two elements x_1 ($\equiv$ satisfactory) and x_2 ($\equiv$ unsatisfactory). Corresponding to each integer θ from 0 to 10 there is a distribution on X and $P_\theta(x_1) = \theta/10$, $P_\theta(x_2) = 1 - \theta/10$. The decision space D has two elements, d_1 ($\equiv$ purchase the ten aircraft) and d_2 ($\equiv$ do not purchase the aircraft). The loss function L is given by the economics of the situation and

$L(d_2, \theta) = r$ for all θ,
$L(d_1, \theta) = r - \theta p + (10-\theta)q$.

The four possible decision functions and their risks are as follows:

(a) $\delta_1 \equiv$ do not purchase whatever the result of the test.

Formally $\quad \delta_1(x_1) = \delta_1(x_2) = d_2$,
and clearly $\quad R_{\delta_1}(\theta) = r$, for all θ.

(b) $\delta_2 \equiv$ purchase whatever the result of the test.

$\quad \delta_2(x_1) = \delta_2(x_2) = d_1$
and $\quad R_{\delta_2}(\theta) = r - \theta p + (10-\theta)q$.

(c) $\delta_3 \equiv$ purchase if the result of the test is satisfactory: do not purchase otherwise.

$\delta_3(x_1) = d_1, \quad \delta_3(x_2) = d_2$.

$$L\{\delta_3(x_1), \theta\} = r - \theta p + (10-\theta)q, \quad (\theta = 1, 2, \ldots, 10)$$
$$L\{\delta_3(x_2), \theta\} = r.$$

Since $P_\theta(x_1) = \dfrac{\theta}{10}$ and $P_\theta(x_2) = 1 - \dfrac{\theta}{10}$,

we have $R_{\delta_3}(\theta) = \dfrac{\theta}{10}\{r - \theta p + (10-\theta)q\} + \left[1 - \dfrac{\theta}{10}\right]r.$

(d) $\delta_4 \equiv$ do not purchase if the result of the test is satisfactory; purchase otherwise.

$$\delta_4(x_1) = d_2, \quad \delta_4(x_2) = d_1.$$

As in (c) we have

$$R_{\delta_4}(\theta) = \dfrac{\theta}{10}r + \left[1 - \dfrac{\theta}{10}\right]\{r - \theta p + (10-\theta)q\}.$$

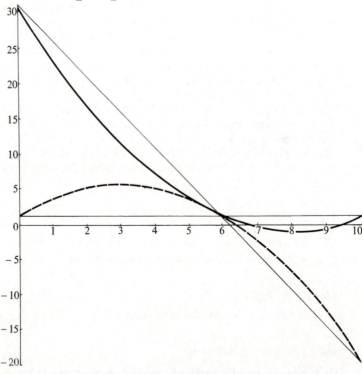

Figure 7 Graphs of risk functions of the four decision functions of Example 11.3.2. Those of δ_1 and δ_2 are straight lines; that of δ_3 the broken curve; and that of δ_4 the unbroken curve

For the sake of illustration, let us assume that $r = 1, p = 2$ and $q = 3$.

Then $R_{\delta_1}(\theta) \equiv 1$,

$R_{\delta_2}(\theta) = 31 - 5\theta$,

$R_{\delta_3}(\theta) = 1 + 3\theta - \frac{1}{2}\theta^2$,

and $R_{\delta_4}(\theta) = 31 - 8\theta + \frac{1}{2}\theta^2$.

The graphs of these functions for $0 \leq \theta \leq 10$ are sketched in Figure 7. (Of course only integral values of θ are relevant.)

As suggested above, there is not a decision function having uniformly minimum risk in this case.

11.4 Minimax decision functions

11.4.1 Wald (1950) has drawn an analogy between the problem of choosing a decision function and the problem of choice of strategy which faces a player in a game against an opponent, who is sagacious and is attempting to win. From this analogy there emerges a criterion for using the risk function to define a 'good' decision procedure. A *minimax* decision function δ is one such that

$$\max_\theta R_\delta(\theta) \leq \max_\theta R_{\delta'}(\theta),$$

for every other decision function δ'. So, for instance, in the above example, δ_1, the procedure according to which the airline does not take up the option whatever the result of the tests, is minimax, since

$$\max_\theta R_{\delta_1}(\theta) = 1 \quad \text{and} \quad \max_\theta R_{\delta_i}(\theta) > 1 \quad \text{for } i = 2, 3, 4.$$

As is illustrated by this example the minimax criterion corresponds to extremely cautious behaviour on the part of the statistician and this is not surprising since the analogy referred to above assigns to Nature the role of the opposing player and thereby implies that Nature is constantly trying to beat the statistician. For this kind of reason, the minimax criterion is not much used in practice.

11.4.2 Another possibility for overcoming the difficulty of the non-existence of uniformly minimum risk procedures, is to impose some 'natural' restriction on the class of decision functions which we shall consider and to hope that within this restricted class there will exist one of uniformly minimum risk.

This of course is exactly what we did in estimation theory when we introduced the condition of unbiasedness, and the problems that beset us there continue to arise here. Sometimes there is no obvious natural restriction to introduce, and even if there is, we have no guarantee that there will exist a uniformly minimum risk procedure in the restricted class. So while this approach may be used for particular problems, it is not of general applicability.

11.5 Admissible decision functions

A third possibility is to use the risk function in a less positive way; rather than think in terms of using it in some way to determine a best procedure, we think in terms of using it only to eliminate bad procedures. This leads to the important notion of admissibility.

We say that a procedure δ strictly dominates a procedure δ' if $R_\delta(\theta) \leqslant R_{\delta'}(\theta)$, for all θ, and this inequality is strict for some θ. Thus in the above example δ_3 strictly dominates δ_4. We have

$$R_{\delta_3}(\theta) < R_{\delta_4}(\theta) \text{ unless } \theta = 5 \text{ or } 6 \text{ in which case } R_{\delta_3}(\theta) = R_{\delta_4}(\theta).$$

Any procedure which is strictly dominated by another is said to be *inadmissible*; any one which is not strictly dominated by another is *admissible*. In the above example, δ_4 is inadmissible because it is dominated by δ_3, while δ_1, δ_2 and δ_3 are admissible.

It may be argued that this is as far as we can reasonably go in the use of the risk function; that any admissible procedure can be justified and it does not really matter which of these we use. Within the class of admissible procedures the choice of one will be made by each individual according to his attitudes. A very cautious individual will choose the minimax procedure. An optimistic individual may well choose that which minimizes minimum risk, or equivalently maximizes maximum expected gain: in the example above either δ_2 or δ_3 would be such a procedure, for the maximum expected gain of each of these is the same; it occurs when $\theta = 10$ and equals 19 which is greater than the maximum expected gain of either δ_1 or δ_4. (It is to be noted, however, in that example, that if the observation made is 'unsatisfactory' then we know that θ is not equal to ten, and it then becomes somewhat absurd to choose a decision procedure on the grounds that this is best when $\theta = 10$.)

11.6 Bayes's solutions

11.6.1 Just as with inference, if we admit a prior distribution on the parameter space which expresses our degrees of belief in different parameter values before an observation is made, then difficulties tend to disappear from the decision problem. For then we may calculate the expected risk $\bar{R}_\delta$, relative to this prior distribution, of any decision function δ and it becomes natural to define as best that function with minimum expected or Bayes's risk. (This is 'rational', as well as natural. Its rationality is best explained in terms of utility. For a discussion of this, see Ferguson, 1967, Section 1.4.) Such a function is called a Bayes's solution to the decision problem.

For instance in the above example on an airline, suppose that our prior degrees of belief in different values of θ are expressed by the statement that all values of θ (the number of good aircraft) between 0 and 10 are equally likely,

i.e., that $P\{\theta = r\} = \dfrac{1}{11}$ $(r = 0, 1, 2, \ldots, 10)$.

Relative to this distribution, $E(\theta) = 5$ and $E(\theta^2) = 35$. From this it follows that

$\bar{R}_{\delta_1} = 1$,
$\bar{R}_{\delta_2} = E[R_{\delta_2}(\theta)] = E(31 - 5\theta) = 6$,
$\bar{R}_{\delta_3} = -1\cdot 5$,
$\bar{R}_{\delta_4} = 8\cdot 5$,

and δ_3 emerges at the Bayes's solution relative to this prior distribution.

Of course relative to another prior distribution for θ, another decision function might have minimum expected risk. It will be seen from Figure 7, that δ_1 has smallest risk for values of θ between 0 and 6, so that δ_1 certainly has minimum expected risk relative to any prior distribution which assigns probability 1 to values of θ between 0 and 6; this may be taken as a formal way of expressing the obvious advice that the airline should not purchase the aircraft if they are convinced beforehand that the number of serviceable aircraft is small. So the Bayesian approach automatically incorporates individual idiosyncrasies through the choice of prior distribution, and in this choice these idiosyncrasies are made quite explicit.

11.6.2 *Computation of Bayes's solutions*

In the above example it is relatively easy to determine the Bayes's solution relative to any given prior distribution simply by calculating the expected risk of each of the four possible decision functions. However in many decision problems – that of estimation, for instance – the class of decision functions is not finite and this elementary approach will not be possible.

The basic problem is apparently one requiring calculus of variations, for we have to choose a function, δ, in order to minimize a functional $\bar{R}$ of it. However we can convert the problem by the following simple expedient.

Suppose that the prior distribution on the parameter space is Π and has density function π with respect to some measure whose element is denoted by $d\theta$. Then an observation x in the sample space converts this prior distribution into a posterior distribution, with density function $\pi(\cdot|x)$, say, according to Bayes's rule. We may then define the expected posterior loss associated with a fixed decision d by

$$E_{\pi(\cdot|x)}\{L(d, \theta)\} = \int_\Theta L(d, \theta)\,\pi(\theta|x)\,d\theta.$$

We can now choose d to minimize this expected posterior loss, for fixed x, and this is *not* a calculus of variations problem; it is a problem of ordinary calculus – determining an element in a space in order to minimize a real-valued function defined on the space. The point of this discussion is that in order to find a Bayes's solution to a decision problem we take $\delta(x)$, for each fixed x, to be that decision d which minimizes the expected posterior loss corresponding to this x.

This follows from

$$\bar{R}_\delta = \int_\Theta \pi(\theta)\,d\theta \int_X L[\delta(x), \theta]\,p(x|\theta)\,dx \qquad \textbf{11.1}$$

$$= \int_X p(x)\,dx \int_\Theta L[\delta(x), \theta]\,\pi(\theta|x)\,d\theta, \qquad \textbf{11.2}$$

since $\pi(\theta)\,p(x|\theta) = p(x, \theta) = p(x)\,\pi(\theta|x)$.

Because $p(x)$ is non-negative we minimize the double integral, that is, $\bar{R}_\delta$, by minimizing the inner integral for each fixed x. In other words we choose $\delta(x)$ for fixed x to be that decision which minimizes the expected posterior loss, and so we may determine the Bayes's solution δ, point by point.

It is worth noting that for fixed x, $\pi(\theta|x) \propto \pi(\theta)\,p(x|\theta)$, so that we may determine the value $\delta(x)$ of the Bayes's solution by minimizing

$$\int_\Theta L[\delta(x), \theta]\,\pi(\theta)\,p(x|\theta)\,d\theta.$$

Evaluation of the 'marginal distribution' of x is not necessary in calculating $\delta(x)$.

11.6.3 Example

With quadratic loss and a uniform prior distribution for the probability θ of success, determine the Bayes's solution to the problem of estimating θ on the basis of the results of n independent trials in which x successes occur.

Here we have

$$\pi(\theta) = 1 \quad (0 < \theta < 1),$$

$$p(x|\theta) = \binom{n}{x}\theta^x(1-\theta)^{n-x},$$

$$L\{\delta(x), \theta\} = \{\delta(x) - \theta\}^2.$$

In an estimation problem such as this it is customary to denote a decision

function by $\hat{\theta}$ rather than by δ. If we do so, then of course,
$$L\{\hat{\theta}(x), \theta\} = \{\hat{\theta}(x) - \theta\}^2.$$
We must choose $\hat{\theta}(x)$ to minimize
$$\int_0^1 \{\hat{\theta}(x) - \theta\}^2 \binom{n}{x} \theta^x (1-\theta)^{n-x} \, d\theta.$$
Let $y = \hat{\theta}(x)$. Then we must choose y to minimize
$$\int_0^1 (y-\theta)^2 \theta^x (1-\theta)^{n-x} \, d\theta.$$
This is simply a calculus problem. Differentiating with respect to y shows that the minimizing value satisfies
$$\int_0^1 (y-\theta)\theta^x (1-\theta)^{n-x} \, d\theta = 0,$$
i.e. $\quad y = \dfrac{\int_0^1 \theta^{x+1}(1-\theta)^{n-x} \, d\theta}{\int_0^1 \theta^x (1-\theta)^{n-x} \, d\theta} = \dfrac{x+1}{n+2}.$

Hence the Bayes's estimator $\hat{\theta}$ is defined by
$$\hat{\theta}(x) = \frac{x+1}{n+2}.$$

Of course $\hat{\theta}(x)$ is just the mean of the posterior distribution of θ given x. Generally, if u is a random variable, we minimize $E(u-y)^2$ by taking $y = E(u)$, and in particular if θ is a random variable with density function $\pi(\theta|x)$, we minimize $E(\theta - y)^2$ by taking $y = E(\theta)$, the expectation being defined relative to the density $\pi(\theta|x)$.

We may now calculate the Bayes's risk of this estimator using either of the forms **11.1** or **11.2**. In this case it is easier to use **11.1**. The inner integral is
$$E_\theta \left[\frac{x+1}{n+2} - \theta \right]^2 = \frac{1}{(n+2)^2} E_\theta [x - n\theta + (1-2\theta)]^2$$
$$= \frac{1}{(n+2)^2} \left[n\theta(1-\theta) + (1-2\theta)^2 \right].$$

Hence the Bayes's risk is
$$\frac{1}{(n+2)^2} \int_0^1 \left[n\theta(1-\theta) + (1-2\theta)^2 \right] d\theta = \frac{1}{6(n+2)}.$$

11.6.4 Example: the two-state, two-decision problem

Suppose that the parameter space Θ has only two elements θ_1 and θ_2 and that the decision space also has two elements, d_1 which corresponds to the decision 'θ_1 is the true parameter' and d_2 likewise for θ_2. Thus we have the problem of testing one simple hypothesis against another.

Let $L_{ij} = L(d_i, \theta_j)$,

so that L_{ij} is the loss incurred by making the decision d_i when θ_j is the true parameter. Usually then we shall have $L_{11} < L_{21}$, because the loss involved in making the correct decision about θ_1 when it is the true state of nature will be less than that involved in making the wrong decision about it; and similarly, $L_{22} < L_{12}$.

Let the prior probabilities of θ_1 and θ_2 be π_1 and π_2 respectively, where $\pi_1 + \pi_2 = 1$, and consider an element x in the sample space. If we take $\delta(x) = d_1$, the expected posterior loss is proportional to

$$\pi_1 p(x|\theta_1) L_{11} + \pi_2 p(x|\theta_2) L_{12}.$$

If we take $\delta(x) = d_2$, the expected posterior loss is proportional (with the same proportionality constant) to

$$\pi_1 p(x|\theta_1) L_{21} + \pi_2 p(x|\theta_2) L_{22}.$$

Hence a Bayes's solution is obtained by taking $\delta(x) = d_1$ or d_2 according as

$$\pi_1 p(x|\theta_1) L_{11} + \pi_2 p(x|\theta_2) L_{12} < \text{or} > \pi_1 p(x|\theta_1) L_{21} + \pi_2 p(x|\theta_2) L_{22},$$

that is, according as

$$\pi_2 p(x|\theta_2) [L_{12} - L_{22}] < \text{or} > \pi_1 p(x|\theta_1) [L_{21} - L_{11}];$$

in other words, assuming $L_{12} > L_{22}$, we take $\delta(x) = d_1$ or d_2 according as

$$\frac{p(x|\theta_2)}{p(x|\theta_1)} < \text{or} > \frac{\pi_1 [L_{21} - L_{11}]}{\pi_2 [L_{12} - L_{22}]}.$$

Thus we are led to a likelihood-ratio test. The set of acceptance of θ_2 is

$$\left\{ x : \frac{p(x|\theta_2)}{p(x|\theta_1)} > k \right\},$$

where $k = \dfrac{\pi_1 [L_{21} - L_{11}]}{\pi_2 [L_{12} - L_{22}]}$.

Note, however, that the value of k is determined by the prior distribution and the loss structure and not by considerations of size. Another way of putting this is to say that the size of the optimal test will vary from problem to problem according to prior degrees of belief and loss structure. It is not surprising really that this deeper analysis of the problem of testing one simple hypothesis against another leads to this result.

One further point is worth mentioning in connexion with this example.

If $\dfrac{p(x|\theta_2)}{p(x|\theta_1)} = \dfrac{\pi_1[L_{21}-L_{11}]}{\pi_2[L_{12}-L_{22}]}$,

what decision do we take?

The answer is that it doesn't matter. For any x satisfying this condition we may take $\delta(x) = d_1$ or $\delta(x) = d_2$ or, for instance, we may toss a coin to decide whether to take $\delta(x) = d_1$ or d_2. In every case we finish with a decision function having minimum Bayes's risk. From a mathematical point of view all that this means is that there may not be a unique decision function having minimum Bayes's risk.

11.6.5 In the previous paragraphs of this chapter we have discussed the simplest version of the decision problem facing the statistician. In particular we have assumed that the informative experiment to be carried out is determined outwith the decision problem. If the choice of experiment is part of this problem its order of difficulty increases considerably. Between the simple version and the version which we have discussed is the case where a sequence of informative experiments is predetermined, where experimentation is costly and where part of the decision problem is to determine when to stop experimentation. We shall discuss below a particular problem of this nature, a so-called Bayes's sequential decision problem, in order to indicate the kind of reasoning required, since this type of problem has received considerable attention in recent years.

We have also avoided certain mathematical questions which arise in decision theory. Does there always exist a Bayes's solution as we have defined it? Are Bayes's solutions always admissible? Is any admissible solution a Bayes's solution relative to some prior distribution? The answers to the latter questions are really rather important. Indeed there is a very close connexion between the class of admissible functions and the class of Bayes's functions, a connexion close enough to provide justification, on purely mathematical grounds, for the use of prior distributions in decision problems. For an excellent account of this and other aspects of decision theory the reader is referred to Ferguson (1967).

11.7 A Bayes's sequential decision problem

On occasion an experimenter has a choice among different experiments which he may conduct before reaching a decision, and he has the problem of which one to choose. The answer is fairly obvious – he should calculate the Bayes's risk associated with each experiment and choose one with minimum Bayes's risk.

For example, consider the airline problem 11.3.2, with $r = 1$, $p = 2$, $q = 3$, and a uniform prior distribution for θ. The airline may have a choice between (i) taking an immediate decision without testing an aircraft, (thereby incurring zero cost for testing), and (ii) testing an aircraft, at the given cost before reaching a decision. We have already calculated the Bayes's risk associated with (ii). It is $-1\cdot5$ (see 11.6.1). It is a simple matter to show that the Bayes's solution to (i) is not to take up the option and that the Bayes's risk is therefore 0. Consequently the airline should choose to test an aircraft before reaching a decision.

Similarly the airline may compare the Bayes's risks associated with the 'experiments' of testing t aircraft, $t = 0, ..., 10$, the cost of testing each aircraft being £1000. However there is another possibility. The airline may have the opportunity to test aircraft sequentially and to choose, at any point, to test another aircraft or to take a decision *on the basis of the results to date*. This raises a much more difficult problem. There are now many more strategies that can be adopted. To discover, at any point, whether it is worth testing another aircraft, it is necessary to compare the Bayes's risk of an immediate decision with that associated with testing another aircraft *and adopting the best strategy thereafter*. This best strategy may be, for instance; if the additional aircraft tested is unsatisfactory, stop testing and decide; if it is satisfactory, test at least one other before deciding.

Now the best future strategy cannot be readily determined by looking ahead. What we have to do is to 'start at the end and work backwards'. We imagine that we have tested all 10 aircraft and found s satisfactory. For each s we obtain the Bayes's solution to the problem of what decision to make, and its risk. (This of course is a very simple matter when all 10 aircraft have been tested, since we are then in the deterministic situation of knowing the value of θ, and risks become known losses.) We enter these risks in a table like Table 1. Now we imagine that we have tested 9 aircraft and found s satisfactory. For illustrative purposes we take $s = 7$. It is an easy matter to calculate the posterior distribution of θ, given the event $A_{7,9}$, that 7 out of 9 tested aircraft have been found satisfactory. This is

$$P(\theta = 7|A_{7,9}) = \frac{3}{11}, \quad P(\theta = 8|A_{7,9}) = \frac{8}{11}.$$

From this we find that the best immediate decision is to take up the option with Bayes's risk 0·4. If, on the other hand, we envisage testing the last aircraft, then with probability 3/11 we incur a loss of 5 and with probability 8/11 zero loss, and so a risk of 1·4. So, given $A_{7,9}$, it is best to stop and take up the option. We enter 0·4 in the appropriate position in Table 1, and then carry out the same calculation for each s from 0 to 9.

Now work back through the table in this way, entering at each stage the risk associated with the best strategy at that stage. Consider for instance the position when 3 aircraft have been tested and 2 found satisfactory. Elementary calculations show that $E(\theta|A_{2,3}) = 6\cdot2$ and that the best immediate decision is to take up the option with a Bayes's risk of 2. If we envisage testing another aircraft we arrive at the event $A_{2,4}$ with a conditional probability, given $A_{2,3}$, of 0·4 and at $A_{3,4}$ of 0·6. Thus the risk associated with testing another aircraft, and adopting the best policy thereafter (which happens to be stop and decide, whatever the result) is 1·0. This is smaller than the Bayes's risk of an immediate decision at $A_{2,3}$, and explains the entry at (2,3) in Table 1.

Table 1. Bayes's risks of optimal strategies in aircraft sequential testing.

Number tested	0	1	2	3	4	5	6	7	8	9	10
Number found satisfactory											
0	−1·8*	(1)	(2)	(3)	(4)	(5)	(6)	(7)	(8)	(9)	(10)
1		−4·7*	(2)†	(3)	(4)	(5)	(6)	(7)	(8)	(9)	(10)
2			−8	1·0*	(4)	(5)	(6)	(7)	(8)	(9)	(10)
3				−10	−1	(5)	(6)	(7)	(8)	(9)	(10)
4					−11	−2·8	3·5	(7)	(8)	(9)	(10)
5						−11·4	−4·0	2·0	7	(9)	(10)
6							−11·5	−4·7	1	5·8	(10)
7								−11·3	−5	0·4	5
8									−11	−5·1	0
9										−10·6	−5
10											−10

* Test another aircraft.
† Either stop or test another.
( ) Having stopped, do not take up the option.

Table 1 shows that there are two optimal strategies, namely: Test one aircraft. If this test is unsatisfactory stop and decide not to take up the option. If it is satisfactory test another. If the second test is satisfactory stop and decide to take up the option. If it is unsatisfactory then either (i) stop and do not take up the option or (ii) test a third aircraft. In case (ii) if the third test is unsatisfactory stop and do not take up the option. If it is satisfactory test a fourth aircraft. If the fourth test is unsatisfactory stop and do not take up the option; if it is satisfactory stop and take up the option.

This is an example of a Bayes's sequential decision problem. The argument is basically that used in dynamic programming and in optimal control theory (see, for instance, Aoki, 1967). It is not immediately clear how this argument can be adapted if there is no upper limit to the number of experiments that can be performed, and for discussion of this see Ferguson (1967) Chapter 7.

Examples

11.1 Let $x_1, x_2, \ldots, x_n$ be a random sample from an $N(\mu, \sigma^2)$ distribution with μ and σ^2 unknown, and suppose that according to a prior distribution μ and $\log \sigma$ are independent and uniformly distributed. Show that, with quadratic loss, $L(\hat{\mu}, \mu) = (\hat{\mu} - \mu)^2$, the Bayes's estimator of μ is $\bar{x}$. What is the Bayes's risk of this estimator?

11.2 The random variable x is $N(\theta, 1)$ and the loss incurred in using an estimate $\hat{\theta}(x)$ for θ is

$$a\{\hat{\theta}(x) - \theta\} \quad \text{if } \hat{\theta}(x) \geq \theta,$$
$$b\{\theta - \hat{\theta}(x)\} \quad \text{if } \hat{\theta}(x) < \theta,$$

where $a > 0$ and $b > 0$. Show that the risk involved in using the estimator $\hat{\theta}_k$ defined by

$$\hat{\theta}_k(x) = x - k$$

can be expressed in the form

$$(a+b)\{\phi(k) + k\Phi(k)\} - ka,$$

where ϕ and Φ are respectively the density and distribution functions of the $N(0, 1)$ distribution. Then show that within the class $\{\hat{\theta}_k : k \text{ real}\}$ an estimator with uniformly minimum risk exists and that the minimum risk is

$$(a+b)\phi\left[\Phi^{-1}\left(\frac{a}{a+b}\right)\right].$$

11.3 Loaves of bread must be at least w grams in weight. At a certain bakery the weight of a loaf from a large ovenload is an $N(\mu, 1/\tau)$ random variable. The parameter μ is a controllable setting, but τ varies from ovenload to ovenload according to a χ^2-distribution with ν degrees of freedom. The cost per loaf of producing an ovenload of mean weight μ is $k + l\mu$, and the selling price of a loaf of satisfactory weight is m. There is no return from an underweight loaf. Show that for ovenloads with setting μ, the mean profit per loaf is

$$m\Psi\{(\mu - w)\sqrt{\nu}\} - (k + l\mu)$$

where Ψ is the distribution function of a t-distribution with ν degrees of freedom. Hence find the best setting μ.

11.4 In measuring the RNA content of a certain type of cell there is difficulty caused by the fact that two cells may be so close together that they are indistinguishable. Two independent measurements x_1 and x_2 are made, each of which may be an observation of the content of one or two (independent) cells. A decision has to be made as to whether the two measurements correspond to two single cells, one single and one pair or two pairs.

Suppose that the distribution of the content of individual cells is known to have density function $xe^{-x}(x \geq 0)$ when measured in suitable units, and there is prior probability π that an observation will derive from two cells rather than one. Suppose also that there is zero loss for a correct decision and unit loss for any wrong decision. Demonstrate by a sketch in the (x_1, x_2) plane, the Bayes's solution of this decision problem.

11.5 A series of n consecutive observations $x_1, x_2, \ldots, x_n$ is made of intensity in a radio receiver. If a signal has been sent then these will be of the form $x_j = a_j + \varepsilon_j$ where $a_1, a_2, \ldots, a_n$ is a known sequence and the random disturbances $\varepsilon_1, \varepsilon_2, \ldots, \varepsilon_n$ have a multivariate normal distribution with zero means and covariance matrix V. If there is no signal then $x_j = \varepsilon_j (j = 1, 2, \ldots, n)$.

One wishes to decide whether a signal has in fact been sent. The prior probability of a signal is p. The losses if one decides mistakenly that a signal has, or has not, been sent are L_1 and L_2 respectively.

What is the decision rule which gives minimum expected loss?

Suppose that this optimum decision rule is being used, and $pL_2 = (1-p)L_1$. What is then the optimum signal sequence $\{a_j\}$ if considerations of transmission power enforce the restriction $\sum a_j^2 = 1$? (*Camb. Dip.*)

11.6 To decide which of two varieties of wheat is to be grown on a large scale in a certain district, an experiment is performed in which n plots are assigned to each variety. The yields $x_{ij} (i = 1, 2; j = 1, 2, \ldots, n)$ are independently and normally distributed with means μ_i and variance σ^2 and μ_1 and μ_2 are taken to have independent normal prior distributions with means α_1 and α_2 and variance σ_0^2. Assuming that σ^2 is known, determine the posterior distribution of (μ_1, μ_2). Deduce that when the loss functions are $L_i = -k\mu_i (i = 1, 2)$, the expected risk is minimized by selecting variety 1 or variety 2 according as

$$x_{1.} - x_{2.} > c \quad \text{or} \quad x_{1.} - x_{2.} < c,$$

where $nx_{i.} = \sum_j x_{ij}$ and $c = (\alpha_2 - \alpha_1)\sigma^2/(n\sigma_0^2)$. (*Camb. Dip.*)

11.7 Observations are to be made sequentially on a process (x_n) of independent Poisson random variables having common mean θ, with the object of estimating θ. Suppose that prior knowledge of θ is described by the improper distribution on $(0, \infty)$ with constant density; that the loss involved in estimating θ by $\hat{\theta}$ is $(\hat{\theta} - \theta)^2$; and that the cost of each observation is c. Find the optimum Bayes's sequential procedure.

11.8 Let $x_1, x_2, \ldots, x_{10}$ be a random sample from an $N(0, \theta)$ distribution. If the prior distribution of θ is that of a $\theta_0 v_0/\chi^2(v_0)$ random variable, where θ_0 and v_0 are positive constants, what is its posterior distribution?

Find the best Bayes's point estimate $\hat{\theta}$ corresponding to the loss function $(\hat{\theta} - \theta)^2$. Prove that as $v_0 \to 0$ while θ_0 remains fixed,

$$\hat{\theta} \to \tfrac{1}{8} \sum x_i^2.$$

Show that the estimator $\tfrac{1}{8} \sum x_i^2$ is inadmissible by comparing it with a suitably chosen multiple of itself. (*Camb. Dip.*)

Appendix A Some Matrix Results

It is assumed that the reader is familiar with linear spaces, Euclidean spaces and the notion of orthogonal subspaces; with linear transformations and matrix representations of these; with range spaces and null-spaces, rank and nullity. For these ideas see Hohn (1964). Here we shall gather together some particular results used in Chapter 3, which are scattered throughout the literature. In what follows, *we shall regard matrices as representations, with respect to fixed orthonormal bases, of linear transformations from one Euclidean space to another.* For the sake of brevity, we shall exploit orthogonality arguments.

A.1   *The range of any matrix* A *is the orthogonal complement of the null-space of* A'

Proof. Let x be a vector in the range of A, so that x is a linear combination of the columns of A; and let y be a vector in the null-space of A', so that $A'y = 0$.

y is orthogonal to each row of A', that is, to each column of A. Hence it is orthogonal to x. Therefore the range of A and the null-space of A', are orthogonal.

Now suppose that A has order $n \times p$, and rank r. The range of A is a subspace of R^n having dimension r. The null-space of A' has dimension $n - \text{rank } A = n - r$, since rank A' = rank A. Therefore the sum of the dimensions of the null-space of A' and the range of A is n. Hence these subspaces are orthogonal complements.

A.2   rank $A'A$ = rank A' [= rank A]

Proof. Let A have order $n \times p$ and let x be any vector in R^n. By section A.1, x is uniquely expressible in the form $x = x_1 + x_2$, where x_1 is in range A and x_2 is in null-space A'. Hence $A'x = A'x_1$.

That is, every vector in the range of A' is in the range of the restriction of A' to the range of A, so that range $A' \subset$ range $A'A$. Clearly range $A'A \subset$ range A'. These two ranges therefore coincide, and rank $A'A$ = rank A'.

A.3   *The equations* $A'A\beta = A'x$ *have a solution for every vector* x

Proof. This follows immediately from the fact that the ranges of $A'A$ and A' coincide, so that, for every x, $A'x$ is in the range of $A'A$.

A.4 *Any solution $\hat{\beta}$ of the equations $A'A\hat{\beta} = A'x$ minimizes $(x - A\beta)'(x - A\beta)$*

Proof. $A'A\hat{\beta} = A'x$ is equivalent to $A'(x - A\hat{\beta}) = 0$.
Now $(x - A\beta)'(x - A\beta) = [x - A\hat{\beta} + A(\hat{\beta} - \beta)]'[x - A\hat{\beta} + A(\hat{\beta} - \beta)]$
$= (x - A\hat{\beta})'(x - A\hat{\beta}) + (\hat{\beta} - \beta)'A'A(\hat{\beta} - \beta)$,
the cross-products vanishing since $A'(x - A\hat{\beta}) = 0$ and $(x - A\hat{\beta})'A = 0$.
Clearly $(\hat{\beta} - \beta)'A'A(\hat{\beta} - \beta) \geq 0$.
Therefore $(x - A\hat{\beta})'(x - A\hat{\beta}) \leq (x - A\beta)'(x - A\beta)$.

A.5 *A symmetric idempotent matrix B of order n represents an orthogonal projection in R^n*

Proof. Because B is idempotent it represents a projection onto its range.
Since $B' = B$ it follows from section A.1 that the range and null-space of B are orthogonal complements. Hence B represents a projection along the orthogonal complement of its range. That is, it represents an orthogonal projection.

A.6 *If A has order $n \times p$ and rank p, the matrix $A(A'A)^{-1}A'$ represents the orthogonal projection of R^n onto the range of A*

Proof. $A(A'A)^{-1}A'$ is symmetric and idempotent and therefore, by section A.5, represents an orthogonal projection. Since $A(A'A)^{-1}A'A = A$, the ranges of $A(A'A)^{-1}A'$ and A coincide.

A.7 *If A has order $n \times p$ and rank p, and x is any vector in R^n, then the square of the distance of x from the range of A is $x'[I - A(A'A)^{-1}A']x$*

Proof. $x = A(A'A)^{-1}A'x + [I - A(A'A)^{-1}A']x$.
The first vector on the right hand side is, by section A.6, the orthogonal projection of x on the range of A. Therefore the distance of x from the range of A is the length of the second vector on the right hand side, and the result follows because $I - A(A'A)^{-1}A'$ is symmetric and idempotent.

A.8 *If Σ is a symmetric positive-definite matrix of order p and H is a $p \times q$ matrix of rank q (so that $q \leq p$), then the partitioned symmetric matrix*

$$\begin{bmatrix} \Sigma & H \\ H' & O \end{bmatrix}$$

is non-singular. (Here O is the zero matrix of order $q \times q$)

Proof. We prove this by construction. We must find a symmetric matrix

$$\begin{bmatrix} P & Q \\ Q & R \end{bmatrix}$$

such that
$$\begin{bmatrix} \Sigma & H \\ H' & O \end{bmatrix} \begin{bmatrix} P & Q \\ Q' & R \end{bmatrix} = \begin{bmatrix} I & O \\ O & I \end{bmatrix}.$$

This requires

$\Sigma P + HQ' = I,$ **A.8.1**

$\Sigma Q + HR = O,$ **A.8.2**

$H'P = O,$ **A.8.3**

$H'Q = I.$ **A.8.4**

From equation **A.8.1** we have $P + \Sigma^{-1}HQ' = \Sigma^{-1}$,

and multiplying this by H' we obtain, using equation **A.8.3**,

$H'\Sigma^{-1}HQ' = H'\Sigma^{-1}.$

Now, since Σ is positive definite and H has full rank, $H'\Sigma^{-1}H$ is non-singular.

Therefore $Q' = (H'\Sigma^{-1}H)^{-1}H'\Sigma^{-1},$
and consequently $P = \Sigma^{-1} - \Sigma^{-1}H(H'\Sigma^{-1}H)^{-1}H'\Sigma^{-1}.$

Also, from equation **A.8.2** $\Sigma^{-1}HR = -Q.$
and, using equation **A.8.4** we obtain $H'\Sigma^{-1}HR = -I.$
Therefore $R = -(H'\Sigma^{-1}H)^{-1}.$

It may now be verified that these values of P, Q and R do in fact provide an inverse for the given matrix.

Appendix B The Linear Hypothesis

There is an application of the likelihood-ratio test which merits separate discussion since it deals with a type of question which arises frequently in practice. We return to the linear model, discussed in chapter 3,

namely $x = A\beta + \varepsilon$,

where x is an n-vector of observations, A is a known $n \times s$ matrix, β is an unknown s-vector and ε is an error vector. In chapter 3 we were interested in estimating β. Here the emphasis changes and we wish to test linear hypotheses concerning β, hypothesis of the form, 'The unknown parameter β satisfies certain linear restrictions, say $H\beta = 0$, where H is a given $r \times s$ matrix.'

This type of problem arises obviously in the context of regression. To take a very simple example, we may have reason to assume that, apart from error, a variable x is linearly related to a single concomitant variable a and we wish to know whether or not 'a really influences x'. Then for observations (x_1, a_1), $(x_2, a_2), \ldots, (x_n, a_n)$, where the a_is are exact measurements and the x_is are subject to error, we have the model

$x_i = \beta_0 + \beta_1 a_i + \varepsilon_i \quad (i = 1, 2, \ldots, n)$

and we wish to test whether $\beta_1 = 0$. (In the above general notation

$A' = \begin{bmatrix} 1 & 1 & \ldots & 1 \\ a_1 & a_2 & \ldots & a_n \end{bmatrix}$

and $H = \begin{bmatrix} 0 & 1 \end{bmatrix}$.)

Similarly we may be prepared to assume that the regression of x on a is quadratic and wish to test whether or not the coefficient of a^2 in the regression equation is zero – in other words whether there is anything to be gained from using quadratic rather than linear regression in our model. The reader may easily visualize the great variety of possible questions when a regression model contains several concomitant variables.

Less obviously, perhaps, the above type of problem occurs when the matrix A can be regarded as a 'design' matrix whose elements are all either 0 or 1. As a simple illustration, suppose that we have r different treatments which may be applied to experimental units and we are prepared to assume that the only possible effects of these treatments are that they shift to varying extents the mean of an underlying distribution of some numerical characteristic x of experimental units. To investigate whether or not the treatments do differ in this way, each treatment is applied to m randomly chosen units and the resulting value of x is observed for each. Let the results be as follows:

Treatment	Observation
1	$x_{11}, x_{12}, \ldots, x_{1m}$
2	$x_{21}, x_{22}, \ldots, x_{2m}$
$\vdots$	$\vdots$
r	$x_{r1}, x_{r2}, \ldots, x_{rm}$.

A model for these observations which incorporates the previous assumptions is

$$x_{ij} = \mu_i + \varepsilon_{ij} \quad (j = 1, 2, \ldots, m; i = 1, 2, \ldots, r),$$

where the ε_{ij}s are independent identically distributed random variables. We are interested in testing the hypothesis that $\mu_1 = \mu_2 = \ldots = \mu_r$. This model and hypothesis fit into the above general framework if we write

$$x' = (x_{11} x_{12} \ldots x_{1m}, x_{21} \ldots x_{2m}, \ldots, x_{r1} \ldots x_{rm}),$$
$$\beta' = (\mu_1, \mu_2, \ldots, \mu_r),$$
$$A = \begin{bmatrix} 1_m & 0 & \ldots & 0 \\ 0 & 1_m & \ldots & 0 \\ \vdots & \vdots & & \vdots \\ 0 & 0 & \ldots & 1_m \end{bmatrix},$$

where 1_m is the m-vector each of whose elements is unity,

and $H = \begin{bmatrix} 1 & -1 & 0 & \ldots & 0 \\ 1 & 0 & -1 & \ldots & 0 \\ \vdots & \vdots & \vdots & & \vdots \\ 1 & 0 & 0 & \ldots & -1 \end{bmatrix}$ of order $(r-1) \times r$.

Again there is a great wealth of possibilities when treatment combinations, complicated designs and complex questions are allowed.

B.1 Returning now to the general model

$$x = A\beta + \varepsilon,$$

we wish to test the hypothesis that $H\beta = 0$. In order to apply the likelihood-ratio test, we must make assumptions about the error vector ε and the assumption which we make is that ε is $N(0, \sigma^2 I_n)$, where σ^2 is unknown, that is, errors are independent *normal* random variables each with zero mean and unknown variance σ^2. Note that this assumption is stronger than that which we adopted in discussing least-squares estimation of β; there we did not assume normality of the error vector. Our present assumption implies that the observation vector x is $N(A\beta, \sigma^2 I_n)$, for some β and some σ^2.

The sample space here is R^n, the labelling parameter θ of the family of possible distributions on the sample space takes the form

$$\theta = (\beta, \sigma^2)$$

and the distribution P_θ on $\mathbf{R}^n$ is defined by the density function

$$p(x, \theta) = \frac{1}{\sigma^n (2\pi)^{\frac{1}{2}n}} \exp\left[-\frac{(x - A\beta)'(x - A\beta)}{2\sigma^2}\right].$$

The null-hypothesis ω is specified by

$$\omega = \{\theta : H\beta = 0\}.$$

Let $\hat\theta = (\hat\beta, \hat\sigma^2)$ be the unrestricted maximum-likelihood estimator of θ and $\dot\theta = (\dot\beta, \dot\sigma^2)$ be the restricted M.L.E. – restricted by the condition $H\beta = 0$. Then, as usual, the critical region of a likelihood-ratio test of ω takes the form

$$\{x : \Lambda(x) > k\},$$

where $\Lambda(x) = \dfrac{p\{x, \hat\theta(x)\}}{p\{x, \dot\theta(x)\}}.$

We may now proceed in a purely formal way to calculate $\hat\theta(x)$, $\dot\theta(x)$ and $\Lambda(x)$. Note that *the assumption of normality which we have made ensures that maximum-likelihood estimates of β are also least-squares estimates.* As the reader may easily verify, the M.L.E., $\hat\beta$, which is obtained by solving the likelihood equations, is that value of β which minimizes $(x - A\beta)'(x - A\beta)$; and $\hat\sigma^2$, again obtained by solving the likelihood equations is

$$\hat\sigma^2 = \frac{1}{n}(x - A\hat\beta)'(x - A\hat\beta).$$

Similarly, $\dot\beta$ minimizes $(x - A\beta)'(x - A\beta)$ subject to $H\beta = 0$

and $\dot\sigma^2 = \dfrac{1}{n}(x - A\dot\beta)'(x - A\dot\beta).$

Therefore $\Lambda(x) = \left[\dfrac{\dot\sigma^2}{\hat\sigma^2}\right]^{\frac{1}{2}n}$

$$= \left[\frac{(x - A\dot\beta)'(x - A\dot\beta)}{(x - A\hat\beta)'(x - A\hat\beta)}\right]^{\frac{1}{2}n},$$

and $\Lambda(x) > k$

is equivalent to $\dfrac{(x - A\dot\beta)'(x - A\dot\beta)}{(x - A\hat\beta)'(x - A\hat\beta)} > k,$

or to $\dfrac{(x - A\dot\beta)'(x - A\dot\beta) - (x - A\hat\beta)'(x - A\hat\beta)}{(x - A\hat\beta)'(x - A\hat\beta)} >$ some constant c.

If we denote by R_1 the residual sum of squares under the null hypothesis ω namely $(x - A\dot\beta)'(x - A\dot\beta)$, and by R_0 the residual sum of squares under the *general model*, namely $(x-A\hat{\beta})'(x-A\hat{\beta})$, then the shape of the critical region of a likelihood ratio test of ω is determined by

$$\frac{R_1 - R_0}{R_0} > c.$$

The problem of determining c to achieve a predetermined size for the test is most easily solved by transforming the original problem to canonical form by an orthogonal transformation of R^n. Basically the problem is this: we know that the mean of a normal distribution on R^n lies in a linear subspace Ω – the subspace consisting of those vectors expressible in the form $A\beta$; we are interested in determining whether this mean lies in a subspace ω_H of Ω, ω_H being defined as the range of the restriction of A to the null space of H or the set of vectors expressible in the form $A\beta$, where $H\beta = 0$.

Suppose that Ω has dimension s and ω_H dimension $s-r$. Then we may change to a new orthonormal basis in R^n whose first $s-r$ components form a basis of ω_H, and whose first s components form a basis of Ω. If under this change of basis, $x \to y$ and $\varepsilon \to \eta$, then our original model may be written, by a suitable reparametrization from β to γ,

$$y_i = \gamma_i + \eta_i \quad (i = 1, 2, \ldots, s)$$
$$= \eta_i \quad (i = s+1, \ldots, n),$$

and ω_H is the subspace for which $\gamma_{s-r+1} = \gamma_{s-r+2} = \ldots = \gamma_s = 0$.

Since by assumption the ε_is are independent $N(0, \sigma^2)$ and since the transformation introduced is orthogonal, the η_is also are independent $N(0, \sigma^2)$. This is the canonical version of the original problem.

It is almost immediately obvious that

$$R_0 = \sum_{i=s+1}^{n} y_i^2$$

and

$$R_1 = \sum_{i=s-r+1}^{n} y_i^2,$$

so that

$$\frac{R_1 - R_0}{R_0} = \frac{y_{s-r+1}^2 + y_{s-r+2}^2 + \ldots + y_s^2}{y_{s+1}^2 + y_{s+2}^2 + \ldots + y_n^2}.$$

When ω is true, $y_{s-r+1}, \ldots, y_n$ are independent $N(0, \sigma^2)$ random variables and so $(R_1 - R_0)/R_0$ is distributed as the ratio of independent $\chi^2_{(r)}$ and $\chi^2_{(n-s)}$ random variables, whatever the true value of σ^2. Therefore, under the null hypothesis

$$\frac{n-s}{r} \frac{R_1 - R_0}{R_0}$$

is distributed as $F_{(r, n-s)}$, and we may determine, simply by reference to tables of the F-distribution, a constant k_α such that

$$P_\theta\left[\frac{n-s}{r}\frac{R_1-R_0}{R_0} > k_\alpha\right] = \alpha \quad \text{for all } \theta \in \omega.$$

In this way a size-α likelihood-ratio test may be constructed and this argument shows this test to be similar.

In practice, of course, we calculate the statistic $(R_1-R_0)/R_0$ by the method first indicated and *not* by finding a transformation which reduces the problem to canonical form. Here the methods introduced in chapter 3 for calculating least-squares estimates and residual sums of squares may be used to advantage. The results of these calculations are often laid out in an 'analysis of variance' table as follows:

	Sum of squares	Degrees of freedom	Mean square	Ratio
For testing ω	R_1-R_0	r	$\frac{1}{r}(R_1-R_0)$	$F = \frac{n-s}{r}\frac{(R_1-R_0)}{R_0}$
Residual	R_0	$n-s$	$\frac{1}{n-s}R_0$	
Total	R_1	$n-s+r$		

The phrase 'for testing ω' is used here to cover different phrases used in different applications. For instance, in the first regression example in this appendix, this phrase might be replaced by 'linear regression of x on a': in the design example, by 'treatments' or by 'differences in treatment means'.

B.2   In practice we are often interested in a more detailed analysis of the variance of observed x_is than that provided, as in the above analysis of variance table, by the test of a single linear hypothesis. For instance, suppose that we have observations $x_1, x_2, \ldots, x_n$, the observation x_i corresponding to values a_i, b_i ($i = 1, 2, \ldots, n$) of two concomitant variables a and b, and suppose that we may adopt the linear regression model

$$x_i = \beta_0 + \beta_1 a_i + \beta_2 b_i + \varepsilon_i \quad (i = 1, 2, \ldots, n),$$

where the εs are independent $N(0, \sigma^2)$ random variables. Then we may be interested in answering the questions: 'Does a influence x?' and 'Does b influence x?' Apparently this means simply that we wish to test the two hypotheses, (a) $\beta_1 = 0$ and (b) $\beta_2 = 0$. However this is not quite so straightforward as it may appear at first sight, since any effects on x of a and b may be impossible to separate on the basis of the observations made. This can be seen quite easily in an extreme case. Suppose that $a_i = b_i$, $i = 1, 2, \ldots, n$. Then the above model may be written in either of the forms

$$x_i = \beta_0 + (\beta_1 + \beta_2)a_i + \varepsilon_i$$
$$\text{or} \quad x_i = \beta_0 + (\beta_1 + \beta_2)b_i + \varepsilon_i.$$

β_1 and β_2 are not identifiable, though $\beta_1+\beta_2$ is. Therefore we cannot hope to test either of the hypotheses (a) and (b), though we may be able to test the single hypothesis that $\beta_1+\beta_2 = 0$. In a less extreme case, where, for instance, a_i is near b_i for each i, separate tests of the hypotheses $\beta_1 = 0$ and $\beta_2 = 0$ may be very poor (have low power) while quite a powerful test of the hypothesis that $\beta_1+\beta_2 = 0$ is possible. Thus we may be able to conclude that at least one of β_1 and β_2 is non-zero without being able to say which is, or whether both are.

This is a general difficulty which occurs when we wish to use a set of data to test more than one hypothesis. In particular, for two hypotheses ω_1 and ω_2 we may be able to conclude that not both are true (to reject $\omega_1 \cap \omega_2$) without being able to carry the analysis further and decide whether it is ω_1 or ω_2 or both which are false. On occasion this further analysis is possible and for *linear* hypotheses there is an orthogonality condition which ensures that it is. This we shall now discuss.

Consider the model

$$x = \phi + \varepsilon$$

where x is an n-vector of observations, ϕ is a mean vector known to lie in a proper subspace Ω of R^n, and ε is an $N(0, \sigma^2 I_n)$ error vector. (This is a slightly more general way of expressing the regression model $x = A\beta + \varepsilon$). Let two linear hypotheses specify respectively that ϕ belongs to the subspaces ω_1 and ω_2 of Ω. Then:

Definition. The linear hypotheses are orthogonal if the orthogonal complements in Ω of ω_1 and ω_2 are orthogonal, that is, if

$$\omega_1^\perp \perp \omega_2^\perp,$$

where $\omega_i^\perp$ denotes the orthogonal complement in Ω of ω_i.

This definition extends in an obvious way to more than two linear hypotheses.

Considerable insight into the theoretical and practical implications of this definition is gained by considering the canonical version of the problem of testing two such hypotheses. We note first that the condition $\omega_1^\perp \perp \omega_2^\perp$ is equivalent to either of the conditions $\omega_2^\perp \subset \omega_1$ or $\omega_1^\perp \subset \omega_2$.

Hence $\dim(\omega_1+\omega_2) \geq \dim(\omega_1+\omega_1^\perp) = \dim \Omega$,

and since ω_1 and ω_2 are subspaces of Ω, it follows that $\dim(\omega_1+\omega_2) = \dim \Omega$. Therefore if the dimensions of $\omega_1 \cap \omega_2$, ω_1, ω_2 and Ω are respectively q, r_1, r_2 and s, the condition $\omega_1^\perp \perp \omega_2^\perp$ implies $s = r_1+r_2-q$. So subject to this condition there exists an orthonormal basis, $e_1, e_2, \ldots, e_n$ of R^n such that

(a) $e_1, e_2, \ldots, e_q$ form a basis of $\omega_1 \cap \omega_2$,
(b) $e_1, e_2, \ldots, e_q, e_{q+1}, \ldots, e_{r_1}$ form a basis of ω_1,
(c) $e_1, e_2, \ldots, e_q, e_{r_1+1}, \ldots, e_s$ form a basis of ω_2,
and (d) $e_1, e_2, \ldots, \ldots\ldots\ldots\ldots, e_s$ form a basis of Ω.

By changing to this orthonormal basis and reparametrizing as previously we obtain the following canonical version of the two orthogonal hypotheses model

$$y_i = \begin{cases} \gamma_i + \eta_i & (i = 1, 2, \ldots, s) \\ \eta_i & (i = s+1, \ldots, n), \end{cases}$$

where the η_is are independent $N(0, \sigma^2)$: the hypotheses that $\phi \in \omega_1$ becomes the hypothesis that $\gamma_{r_1+1} = \gamma_{r_1+2} = \ldots = \gamma_s = 0$; the hypothesis that $\phi \in \omega_2$ becomes $\gamma_{q+1} = \gamma_{q+2} = \ldots = \gamma_{r_1} = 0$; and the hypothesis that $\phi \in \omega_1 \cap \omega_2$ becomes $\gamma_{q+1} = \gamma_{q+2} = \ldots = \gamma_s = 0$.

The sum of squares for testing ω_1 is $y_{r_1+1}^2 + \ldots + y_s^2$; that for testing ω_2 is $y_{q+1}^2 + y_{q+2}^2 + \ldots + y_{r_1}^2$. These are independent and their sum is the sum of squares for testing $\omega_1 \cap \omega_2$.

When the orthogonality condition is satisfied, it is therefore possible to draw up the following analysis-of-variance table:

	Sum of squares	Degrees of freedom	Mean square	Ratio
For testing ω_1 (i)	$R_1 - R_0$	$s - r_1$	$m_1 = \dfrac{R_1 - R_0}{s - r_1}$	$F_1 = \dfrac{m_1}{m_0}$
For testing ω_2 (ii)	$R_2 - R_0$	$s - r_2 = r_1 - q$	$m_2 = \dfrac{R_2 - R_0}{s - r_2}$	$F_2 = \dfrac{m_2}{m_0}$
For testing $\omega_1 \cap \omega_2$ (i)+(ii)	$= R_{12} - R_0$	$s - q$	$m_{12} = \dfrac{R_{12} - R_0}{s - q}$	$F_3 = \dfrac{m_{12}}{m_0}$
Residual	R_0	$n - s$	$m_0 = \dfrac{R_0}{n - s}$	
Total	R_{12}	$n - q$		

In this table R_1, R_2 and R_{12} are the residual sums of squares under the hypotheses $\theta \in \omega_1$, $\theta \in \omega_2$, $\theta \in \omega_1 \cap \omega_2$ respectively.

The orthogonality condition, with its consequent partition of the sum of squares for testing $\omega_1 \cap \omega_2$, represents, in a sense, the opposite extreme from that exemplified above by the case of two hypotheses ($\beta_1 = 0, \beta_2 = 0$) which were not 'separable' because of non-identifiability of β_1 and β_2. The extent to which we may reliably carry a two-hypothesis analysis beyond the stage of deciding whether $\omega_1 \cap \omega_2$ is true, to the stage of deciding about the separate hypotheses, depends on which of these extremes we are nearest. Consequently it is often necessary to take thought before experimenting or collecting data in order to ensure that these data will be reasonably informative regarding the questions which one wishes to answer. In particular it is often desirable to design an experiment which ensures that the orthogonality condition is satisfied for linear hypotheses of interest. The subject of experimental design is a very large one on which many books have been written. The interested reader is referred to Cox (1958), and Cochran and Cox (1950) for an introduction to this subject.

We conclude our rather brief discussion of this important topic by returning to the regression model with two concomitant variables, namely

$$x_i = \beta_0 + \beta_1 a_i + \beta_2 b_i + \varepsilon_i \quad (i = 1, 2, \ldots, n),$$

where we are interested in the hypotheses (a) $\beta_1 = 0$ (b) $\beta_2 = 0$. This can be written in the form

$$x = A\beta + \varepsilon,$$

where $A' = \begin{bmatrix} 1 & 1 & \ldots & 1 \\ a_1 & a_2 & \ldots & a_n \\ b_1 & b_2 & \ldots & b_n \end{bmatrix}$

Thus in the notation of the present section, Ω is the range of the matrix A. Also ω_1, the subspace specified by the hypotheses that $\beta_1 = 0$, is the range of the matrix A_1, where

$$A_1' = \begin{bmatrix} 1 & 1 & \ldots & 1 \\ b_1 & b_2 & \ldots & b_n \end{bmatrix},$$

and ω_2, the subspace specified by the hypothesis that $\beta_2 = 0$ is the range of A_2, where

$$A_2' = \begin{bmatrix} 1 & 1 & \ldots & 1 \\ a_1 & a_2 & \ldots & a_n \end{bmatrix}.$$

So ω_1 is spanned by the vectors u, where $u' = (1, 1 \ldots, 1)$ and b, where $b' = (b_1, b_2, \ldots, b_n)$. Since we are interested in orthogonality, it is more natural to consider ω_1 as being spanned by u and by $b - \bar{b}u$, where $\bar{b} = n^{-1} \sum b_i$, since u and $b - \bar{b}u$ are orthogonal. Similarly ω_2 is spanned by u and $a - \bar{a}u$, while $\omega_1 \cap \omega_2$ is spanned by u. If we assume identifiability of β_1 and β_2, then not all the as are equal, not all the bs are equal and A has rank 3, so that Ω, ω_1 and ω_2 have dimensions 3, 2 and 2 respectively. $\omega_1^{\perp}$ has dimension 1 and is spanned by a vector in Ω orthogonal to both u and $b - \bar{b}u$: this vector is contained in ω_2 if and only if it is of the form $\lambda(a - \bar{a}u)$. Hence the hypotheses $\beta_1 = 0$ and $\beta_2 = 0$ are orthogonal if and only if $b - \bar{b}u$ and $a - \bar{a}u$ are orthogonal, that is, if and only if $\sum (a_i - \bar{a})(b_i - \bar{b}) = 0$.

Having translated the original orthogonality condition into this usable form we can use this example to demonstrate how orthogonality of hypotheses facilitates calculations by obtaining explicit expressions for the sums of squares in an analysis of variance table. Note first that the model may be written as

$$\begin{aligned} x_i &= (\beta_0 + \beta_1 \bar{a} + \beta_2 \bar{b}) + \beta_1 (a_i - \bar{a}) + \beta_2 (b_i - \bar{b}) + \varepsilon_i \\ &= \delta_0 + \beta_1 (a_i - \bar{a}) + \beta_2 (b_i - \bar{b}) + \varepsilon_i, \text{ say.} \end{aligned}$$

Now let $\hat{\delta}_0$, $\hat{\beta}_1$, $\hat{\beta}_2$ be unrestricted maximum-likelihood estimates of δ_0, β_1 and β_2 respectively; and let $\acute{\delta}_0$, $\acute{\beta}_1$ be estimates restricted by the condition $\beta_2 = 0$; $\grave{\delta}_0$ and $\grave{\beta}_2$ estimates restricted by $\beta_1 = 0$. The orthogonality conditions $\sum(a_i - \bar{a}) = \sum(b_i - \bar{b}) = \sum(a_i - \bar{a})(b_i - \bar{b}) = 0$ imply, as is readily verified, that

$$\hat{\delta}_0 = \grave{\delta}_0 = \acute{\delta}_0 = \bar{x},$$
$$\hat{\beta}_1 = \acute{\beta}_1 = \frac{\sum(a_i - \bar{a})x_i}{\sum(a_i - \bar{a})^2},$$
$$\hat{\beta}_2 = \grave{\beta}_2 = \frac{\sum(b_i - \bar{b})x_i}{\sum(b_i - \bar{b})^2}.$$

Explicit expressions for the sums of squares in an analysis-of-variance table are easily obtained and we find that

$$R_{12} = \sum(x_i - \bar{x})^2,$$
$$R_0 = \sum\{x_i - \bar{x} - \hat{\beta}_1(a_i - \bar{a}) - \hat{\beta}_2(b_i - \bar{b})\}^2$$
$$= \sum(x_i - \bar{x})^2 - \hat{\beta}_1^2 \sum(a_i - \bar{a})^2 - \hat{\beta}_2^2 \sum(b_i - \bar{b}),$$
$$R_1 = \sum\{x_i - \bar{x} - \grave{\beta}_2(b_i - \bar{b})\}^2 = \sum(x_i - \bar{x})^2 - \hat{\beta}_2^2 \sum(b_i - \bar{b})^2,$$
$$R_2 = \sum\{x_i - \bar{x} - \acute{\beta}_1(a_i - \bar{a})\}^2 = \sum(x_i - \bar{x})^2 - \hat{\beta}_1^2 \sum(a_i - \bar{a})^2.$$

Thus the analysis-of-variance table reduces to the following:

	Sum of squares	Degrees of freedom
For testing $\beta_1 = 0$	$\hat{\beta}_1^2 \sum(a_i - \bar{a})^2$	1
For testing $\beta_2 = 0$	$\hat{\beta}_2^2 \sum(b_i - \bar{b})^2$	1
For testing $\beta_1 = \beta_2 = 0$	$\hat{\beta}_1^2 \sum(a_i - \bar{a})^2 + \hat{\beta}_2^2 \sum(b_i - \bar{b})^2$	2
Residual	$\sum(x_i - \bar{x})^2 - \hat{\beta}_1^2 \sum(a_i - \bar{a})^2 - \hat{\beta}_2^2 \sum(b_i - \bar{b})^2$	$n - 3$
Total	$\sum(x_i - \bar{x})^2$	$n - 1$

This table has considerable intuitive meaning and might well have been arrived at without our detailed analysis involving the use of the likelihood-ratio test. The point of the detailed analysis is of course that we can use the general results achieved by it in applications where intuition is an insufficient guide. There are many such applications of varying degrees of complexity.

Analysis of variance is a very important practical tool which emerges, as above, from considering the application of the likelihood-ratio method to the problem of testing linear hypotheses. There are other models and questions for which this technique is appropriate and a full discussion of these is given by Scheffé (1960). For these other models there are subtle differences in interpretation and in properties of the tests involved, so that it is dangerous to apply the technique without being quite specific about the model and questions involved in any particular application. The present section is intended to be no more than the briefest of introductions to this important topic.

References

AOKI, M. (1967), *Optimization of Stochastic Systems*, Academic Press.
ARMITAGE, P. (1960), *Sequential Medical Trials*, Blackwell, Oxford.
COCHRAN, W. G., and COX, G. M. (1950), *Experimental Designs*, Wiley.
COX, D. R. (1952), 'Sequential tests for composite hypotheses', *Proc. Camb. Phil. Soc.*, vol. 48, pp. 290–99.
COX, D. R. (1958), *Planning of Experiments*, Wiley.
COX, D. R., and MILLER, H. D. (1965), *The Theory of Stochastic Processes*, Methuen.
CRAMÉR, H. (1937), *Random Variables and Probability Distributions*, Cambridge Tracts in Mathematics, no. 36, Cambridge University Press.
CRAMÉR, H. (1946), *Mathematical Methods of Statistics*, Princeton University Press.
DOOB, J. L. (1953), *Stochastic Processes*, Wiley.
FELLER, W. (1968), *An Introduction to Probability Theory and its Applications*, 3rd edn, Wiley.
FERGUSON, T. S. (1967), *Mathematical Statistics – A Decision Theoretic Approach*, Academic Press.
FINNEY, D. J. (1947), *Probit Analysis*, Cambridge University Press.
HAJEK, J., and SIDAK, Z. (1967), *Theory of Rank Tests*, Academic Press.
HODGES, J. L., and LEHMANN, E. L. (1964), *Basic Concepts of Probability and Statistics*, Holden-Day, San Francisco.
HOHN, F. E. (1964), *Elementary Matrix Algebra,* 2nd edn, Macmillan, New York.
LEHMANN, E. L. (1959), *Testing Statistical Hypotheses*, Wiley.
LEHMANN, E. L., and SCHEFFÉ, H. (1950), 'Completeness, similar regions and unbiased estimation', *Sankhya*, vol. 10, pp. 305–40.
LINDGREN, B. W. (1962), *Statistical Theory*, Macmillan, New York.
LINDLEY, D. V. (1965), *Introduction to Probability and Statistics from a Bayesian Viewpoint* (2 volumes), Cambridge University Press.
MALINVAUD, E. (1966), *Statistical Methods of Econometrics*, North-Holland, Amsterdam.
MEYER, P. L. (1965), *Introductory Probability and Statistical Applications*, Addison-Wesley, Massachusetts.
NOETHER, G. E. (1967), *Elements of Nonparametric Statistics*, Wiley.
SCHEFFÉ, H. (1960), *The Analysis of Variance*, Wiley.
SILVEY, S. D. (1959), 'The Lagrangian multiplier test', *Ann. Math. Statist.* vol. 30, pp. 389–407.
WALD, A. (1943), 'Tests of statistical hypotheses concerning several parameters when the number of observations is large', *Trans. Am. Math. Soc.*, vol. 54, pp. 426–82.
WALD, A. (1947), *Sequential Analysis*, Wiley.

WALD, A. (1949), 'Note on the consistency of the maximum-likelihood estimate', *Ann. Math. Statist.*, vol. 20, pp. 595-601.
WALD, A. (1950), *Statistical Decision Functions*, Wiley.
WALSH, J. E. (1962), *Handbook of Nonparametric Statistics*, Van Nostrand.
WETHERILL, G. B. (1966), *Sequential Methods in Statistics*, Methuen.

Index

Admissibility
  and Bayes's procedures 168
  definition of 166
  example of 163–6
Analysis of variance 180–88
Aoki, M. 171
Armitage, P. 136
Association
  (in contingency table) 94
  likelihood-ratio test for 112, 114–15
Bayesian
  confidence interval 155
  inference, generally 153–60
  inference for hypotheses 155–6
Bayes's risk 166
Bayes's sequential decision procedure 171–4
Bayes's solutions
  admissibility of 168
  computation of 167–8
  definition of 166
  examples of 168–71
Chi-squared test
  generally 118–20
  goodness-of-fit 142
  for multinomial distributions 120
Cochran, W. H. 187
Completeness 29–34
  definition of 29
  of exponential families 31
  and minimal sufficiency 30, 34
  and minimum-variance unbiasedness 33
Composite hypothesis 102
Conditional tests 145–7
Confidence coefficient 88, 89
Confidence ellipsoid 91–2
Confidence interval, example 87–8, 90
  see also Bayesian confidence interval
Confidence set
  definition of 88–9
  construction of 89
  examples of 90–92
  optimal properties of 92
Consistency
  definition of 76
  of maximum likelihood 74–7
Cox, D. R. 132, 135, 187
Cox, G. M. 187
Cramér, H. 78, 91
Cramér-Rao inequality 35–7
  for vector parameter 41–3
Cramér-Rao lower bound 38–9, 73
Critical region, definition of 96

Decision function 162
Decision space 162
Design matrix 180
Design of experiment 123, 186
Distribution-free statistic 141
Dominance 166
Doob, J. L. 29
Efficiency 39–40
  of maximum likelihood 74, 77–8
Empirical distribution function 141
Error probabilities
  of sequential tests 125–7, 130–35
  Type 1 96
  Type 2 96
Estimable functions 53, 54
Estimate. definition of 23
Estimator, definition of 23
Expected posterior loss 167–8
Exponential family 31
Factorization theorem 27
Feller, W. 13
Ferguson, T. S. 167, 171, 174
Finney, D. J. 73
Fisher's information 37, 40
Fundamental lemma (Neyman-Pearson) 98–101
Gauss-Markov theorem 51–4
Hajek, J. 150
Hodges, J. L. 43
Hohn, F. E. 48, 177
Hypothesis, definition of 95
Identifiability, definition of 50
Improper distribution 158
Information, Fisher's 37
Information matrix 41, 71, 78
  example of 73
  and non-identifiability 81–2
Invariant tests 105
Jensen's inequality 29, 75
Kolmogorov-Smirnov test 140–42
Lagrange multipliers 60, 79
Least-squares estimates 46–67
  optimal properties of 51–4, 58–9
  variance of 57
  weighted 54–7
  with side conditions 59–64
Lehmann, E. L. 26, 27, 31, 43, 101, 105, 110, 111, 125, 131, 136, 144, 147
Likelihood function, definition of 68
Likelihood-ratio test 100, 108–9

examples   101, 109–12
for linear hypothesis   182
Likelihood-ratio statistic   109
large-sample distribution of   112–15
Lindgren, B. W.   13
Lindley, D. V.   13, 21, 159
Linear hypotheses   180–88
canonical form of   183
orthogonality of   185–8
Loss function   161, 162

Malinvaud, E.   65
Maximum-likelihood estimates   68–83
calculation of   70–71
consistency of   74–6
definition of   68
large-sample properties of   74–8
optimal properties of   73–4
subject to restrictions   79–82
Mean-square error   24
Meyer, P. L.   13, 29, 36, 38
Miller, H. D.   132
Minimal sufficiency   26, 27, 58
and maximum-likelihood estimates   74
Minimax procedures   165
Monotone likelihood ratio   131
Most-powerful test   98, 100
example of   101, 102

Newton's method   70
Neyman–Pearson
theory   96–106
fundamental lemma   98–101
Noether, G. E.   149
Non-identifiability   50, 62–4, 81–2, 119–20
Non-parametric tests   139–52
Normal equations   47

Order statistic   147
Orthogonal linear hypotheses   185–8

Parameter, definition of   18
Parametric problems, definition of   108
Permutation tests   144–5, 148–50
as conditional tests   147–8
Pivotal quantity   89
Point estimate, definition of   18
Posterior distribution, definition of   154
Power of tests, definition of   96
Prior distribution
definition of   153
choice of   157
Probability density   15
Probability distribution   14
Probit analysis   73

Quadratic loss   162

Randomization   148–50
Randomized test   100–101
Random sample   16

Random variable, definition of   16
Random walk   132, 134
Rao–Blackwell theorem   28, 58
Reparametrization   58, 60, 79, 115
Residual mean-square   90
Residual sum of squares   56, 182
distribution of   59
Restricted estimates   59–63, 79–82
Risk function   162

Sample distribution function   141
Sample space   14
Scheffé, H.   27, 54, 188
Sequential decision procedure (Bayes's)   171–4
Sequential probability ratio test
definition of   125
error probabilities of   125–7, 131–3
examples of   127–9, 133–5
expected numbers of observations for   132–3
graphical procedure for   127–9
optimal properties of   125, 136
Sequential procedure   124
Set estimation   18, 87
Sidak, Z.   150
Significance level, definition of   97
Silvey, S. D.   82, 119
Similar test, definition of   111
Simple hypothesis, definition of   98
Singularity (of information matrix)   81–2
Size of test, definition of   97
Statistic, definition of   26
Sufficiency   25–8
of partition   25
of statistic   27
and test construction   145–7

Test of hypothesis, definition of   94–5
Trial   21

Unbiased estimate   24
of minimum variance   24–35
Unbiased test   104–5
Uniformly minimum risk   162–3, 165
Uniformly most-powerful test
definition of   97
example of   103
existence of   103–4

Variance estimate   56–7, 62
Variance matrix, definition of   51

Wald, A.   77, 135, 165
Wald's Identity   132, 133
Walsh, J. E.   150
Wetherill, G. B.   136, 137
Wilcoxon test   143–4
W-test   115–16
example of   116–18